Introducing Astronomy

by the same author

★

AMATEUR ASTRONOMER'S HANDBOOK

OBSERVATIONAL ASTRONOMY FOR AMATEURS

Introducing Astronomy

by

J. B. SIDGWICK
F.R.A.S.

SECOND EDITION
revised by
R. C. GAMBLE

FABER AND FABER LIMITED
3 Queen Square
London

First published in 1951
by Faber and Faber Limited
3 Queen Square London W.C.1
First published in this edition 1959
Reprinted 1961 and 1968
New and revised edition 1973
Printed in Great Britain by
Whitstable Litho, Straker Brothers Ltd

ISBN 0 571 04823 4

Contents

Author's Note

The present book is a radical revision of my *Astronomy for Night Watchers*, first published fourteen years ago and intended as a guide to the night sky and an outline of the elements of descriptive astronomy. It was hoped that it would help to while away the long sleepless hours of wardens, Observer Corps personnel, members of ambulance and rescue squads, fire watchers, and others to whom the peculiar conditions of that time offered an opportunity of exploring a new field. Both the selection and the presentation of the material were designed with that end in view.

Conditions change. A popular introduction to astronomy can today afford to cover wider ground, and be less superficial in treatment, more general in its appeal. The original text has therefore been largely rewritten, brought up to date in a number of respects, and enlarged in scope—notably by the addition of three new chapters. The revision has, in fact, been sufficiently thoroughgoing to justify its issue under a different title, *Introducing Astronomy*.

J.B.S.

London,
December 1957

Chapter 1

WHY STUDY ASTRONOMY?

Cartoonists, who not only express popular misconceptions but do much to mould them, usually envisage an astronomer as a doddering greybeard, too absent-minded to cope effectively with the demands of everyday life because his head is beyond the clouds, among the stars. Popularly, his work and interests are conceived of as being compounded of strings of differential equations and unintelligible jargon in more or less equal proportions. He is absorbed in something which is meaningless as well as being practically useless.

No picture could, in fact, be more distorted. The fascination of astronomy is there for all who still retain a capacity for appreciating beauty, and whose eyes are not closed to the incredible wonders and fundamental strangeness of the universe we inhabit. Throughout the ages men have gazed up at the stars and wondered. The man has not been born who has never been moved by the immensity of a frosty starlit night, when the skies seem to coruscate with an infinity of flashing stellar points. The man is unborn, too, who has never, if only for a fleeting moment, wished to know something about it all.

I want first of all to emphasize that this instinct is a valid one, and then to show that astronomy is *not* too difficult for the 'average man' to understand and enjoy. On the contrary, the amateur study of astronomy is a pursuit that can never grow stale and which can provide endless hours of absorbing interest, both by day and by night.

Even without any instrumental aid whatsoever, hours may be spent in tracing out the ancient constellation figures, learning their names and their long histories, the myths of classical Greece and Rome with which they are associated. Gradually, in this way, one learns to find one's way about the night sky. No longer is it just a riot of starry points, strange and indistinguishable one from another, as chaotic as a Cup Final crowd seen from an aeroplane. One by one the brighter star groups fall into place, and chaos gives way to order. The sky ceases to be a crowd of miscellaneous aliens and becomes a circle of familiar friends, each with his well-known appearance and personal idiosyncrasies. To see Orion striding over the eastern horizon on an autumn evening after six months' absence below the horizon is to experience the same thrill one feels when, quite unexpectedly, one sees the face of an old friend who for months has been out of sight and out of mind.

If the observer has a pair of binoculars at his disposal a further and even more fascinating field is opened to him. In these days when giant new telescopes costing millions of dollars and weighing many tons have considerable news value, it is often imagined that without expensive equipment and a long technical training the pursuit of astronomy must be a waste of time, and the night skies a closed book. Nothing, once more, could be further from the truth. The amount one is able to see at any time depends, roughly speaking, upon the size of the lens that is gathering light and transmitting it to the retina. Look at your eye in a mirror, note the size of the pupil, and then compare it with the object glass of the smallest telescope or field glass. The difference between the two is a gauge of the new wonders that even the most modest instrument will reveal. To one who has previously relied solely on his unaided eyesight, the first view with binoculars of the crescent moon, or the Milky Way, or the star clusters in the constellation Perseus—the list could be extended indefinitely—is a revelation. The owner of a pair of field glasses holds the open sesame to realms of whose very existence his naked eyes never gave him the faintest inkling.

Amateurs—many of whom began with the smallest instruments the work that later brought them fame—hold a distinguished position in the history of astronomy. Their achievements would fill a book many times the size of this; here there is only space to mention two or three.

One of the most fundamental properties of the sun is a tendency for all its activities and phenomena to fluctuate in intensity in recurrent cycles of some eleven years' duration. This was the discovery of Schwabe, a German amateur (he was an apothecary by profession) who began his observations of the sun with a small and inexpensive telescope in 1826. Every cloudless day for more than forty years he studied the face of the sun with his little telescope, noting how many spots were visible. In 1852 he was able to announce that sunspot activity fluctuates between a definite maximum and minimum in a period of about eleven years. This was one of the most significant discoveries ever made in solar research —the work of an amateur who indulged his astronomical ardour at such times as he could spare from the daily routine of earning a living.

Our knowledge of meteors, and of the groups of swarms in which they are organized, owes more to one amateur than to any other man. W. F. Denning, of Bristol, made his first meteor observation with the naked eye at the age of twenty-one, when he saw a fireball belonging to the Taurid swarm. Thirty years later he published his *General Catalogue*, a standard reference work on meteor radiants. Denning was also one of the foremost planetary observers of his time, and the discoverer of five comets and two novae. Yet he never used a telescope larger than $12\frac{1}{2}$ inches in diameter, simply—even crudely—mounted.

Another branch of astronomy—double stars—likewise owes an enormous debt to the skill and pertinacity of an amateur—Burnham, who until eleven years before his death in 1921 was an official reporter at the U.S. District Court, Chicago. His great *Catalogue* is also one of the source books of astronomy; his discoveries numbered 1,340.

Coming nearer our own times, the reader probably knows that Will Hay, the scholastic stage and screen star, was an accomplished amateur astronomer, and he may recollect that in the early 1930's Hay discovered a bright spot on the surface of Saturn which was one of the most conspicuous markings ever observed on that planet. Another amateur whose love of astronomy has led him into the halls of fame is L. C. Peltier, an American farmer. Working his farm by day, night-time will find him in his home-made observatory, where between 1925 and 1939 he discovered no fewer than six new comets. In an extremely interesting letter to me he describes his equipment and methods of working. The former is simple and inexpensive, and the latter neither esoteric nor complicated: Peltier is a true amateur who has in recent years shown that 200-inch telescopes and the like have not yet elbowed the amateur from the niche he has carved for himself in the history of astronomical discovery. More recently still one may cite the work of the British amateur G. E. D. Alcock, who made history in 1967 with his discovery of a nova in the constellation of Delphinus, working only with powerful binoculars (11 × 80). Mr. Alcock has subsequently repeated this feat.

Innumerable works on descriptive astronomy have been written, giving the armchair astronomer a good idea, couched in popular terms, of what the whole thing is about: what the sun, planets and stars are, their sizes, temperatures and distances, the manner in which the contents of the material universe are organized and arranged. The popularity of such books shows that a great many people are sufficiently interested at any rate to read about astronomy. But the man who confines his astronomy to the armchair is as far from the real thing, the genuine *frisson* that results from direct contact, as one whose experience of music is confined to reading the press notices of concerts. It is interesting, no doubt, to learn that Jupiter has twelve moons, four of which are large enough to have been discovered with Galileo's primitive telescope; but to go out into the garden one night and see them for oneself is the only way to let astronomy exert its full fascination.

Even with nothing more than binoculars, the night sky reveals itself as a treasure house of marvels, invisible to the naked eye, and the only direct method of experiencing what astronomy has to offer is the practical method—descriptive books are excellent and invaluable as far as they go, but their full significance and interest is withheld unless the reader takes the further step of seeing for himself.

A few practical tips may be of value here. First, about binoculars. A large number of first-rate glasses have come into the market since the war, through army disposals firms, and a good pair can be bought quite cheaply.

Prismatic 'field glasses', with object glasses from $1\frac{1}{2}$ to 2 inches diameter, are superior to the smaller variety known as opera glasses. Engraved somewhere on the body you may find some such legend as '7×50'; this means that the glasses enlarge the image seven times, and that the diameter of the object glasses is 50 mm., or about 2 inches.

Determining the magnification of the binoculars for yourself is a simple matter. Look at a brick wall through the left eyepiece with your right eye, keeping your left eye open. The two images of the wall, one formed by the naked eye and one by the optical train of the binoculars, are superimposed, and it is only necessary to count how many bricks of the former coincide with one of the latter to arrive at the magnification: if it is found, for example, that five bricks of the naked eye image coincide with one of the binocular image, then the magnification is five.

Any magnification from $\times 3$ up to $\times 10$ or $\times 12$ is useful: the lower magnifications for giving wide-field views of the constellations, and the higher magnifications for use on such objects as the moon, planets and double stars. With a magnification of $\times 3$, double stars should be separated if they are more than about $1'$ apart;[1] a pair of $\times 12$ binoculars should split double stars whose separation is only about $20''$. A good pair of binoculars of the size mentioned above will show stars down to about the eighth magnitude[2]—roughly three times fainter than the faintest stars visible with the naked eye.

Most good binoculars are made on a hinged principle, so that the distance separating the two tubes can be varied to suit different users. See that they are properly adjusted in this respect or a double image of the field will result and the glasses will be virtually useless. Good glasses also allow for the separate focusing of each eyepiece, since with most people the focus of the two eyes is not exactly the same. It is obviously of the greatest importance that each eyepiece should be focused as nearly perfectly as possible,

[1] See p. 133. [2] See p. 86.

and it is preferable to focus the instrument on a star, rather than on some distant terrestrial object.

A very handy little instrument, recently available through army disposals, is the predictor telescope. This is a 6 × 48 monocular, less than a foot long. With an object glass nearly 2 inches in diameter, and a field of 8°, it gives superb views of the constellations and Milky Way, as well as being admirable for the observation of sunspots, the lunar craters and mountains, comets, etc.

How best to keep warm while observing at night is a matter of practical importance, as anyone will be prepared to agree after a couple of hours spent immobile out of doors on a frosty night, with an east wind whipping round the corner. Do not drink hot tea or coffee just before going out: its heating effect is only transitory and it renders one susceptible to chills on the stomach. The time to take a hot drink is when one comes in. It is an excellent plan to warm one's underclothes in front of a fire before going out. Loose-fitting clothes are better than tight, and the ideal type of garment is a thick woolly pullover or sweater several sizes too large. Wearing two pairs of gloves and trousers also makes a lot of difference: pyjama trousers inside the ordinary pair, and woollen gloves inside leather ones help materially to conserve warmth. Duplicated woollen socks inside loose-fitting boots such as gum boots or waders are a help, but even this will not keep the feet warm for long—the most difficult thing to achieve. The secret is not to stand or sit with the feet resting on the ground itself (or concrete roof or whatever it may be) but on some non-conducting material such as a piece of coco-nut matting or a thick coco-nut fibre doormat.

Lastly, one very important point. In order to catch a glimpse of faint objects such as the companions of some double stars, star clusters and nebulae near the limit of vision, etc., the binoculars must be held perfectly steady. This the hand, alone and unsupported, cannot do: the steadiest hand on earth may be holding the glasses, but it will be found that the star images are dancing about like sparks over a furnace. It is imperative that some sort of support for the glasses, and/or the hands holding them, should be improvised. The best thing is possibly an ordinary step ladder, against the top step of which the hands holding the binoculars can be steadied. Alternatively, use can be made of an easy chair (if the object being observed is not too high in the sky) whose arms give firm support to the elbows. It is also worth noting in passing that the centre of the retina is not its most sensitive area. Therefore it often happens that a faint object is hardly discernible when the attention and the eyes are focused directly at it, while it is easily seen when the eyes are focused at a point a little way from it. In this way—by having a firm support for the glasses and by using indirect vision—faint objects can be detected that would otherwise inevitably remain invisible.

Having thus introduced the subject and cleared the ground of one or two preliminary points, we can set out on our voyage of discovery.

Chapter 2

THE FIRST NIGHT OUT—FINDING ONE'S BEARINGS

The Constellations

One of the first things we notice when we begin our observations on a starlit night is that the brighter stars are not distributed uniformly over the dome of the heavens, but in many cases fall into quite obvious groups. These groups are the constellations, and their identification will be the main concern of the beginner, for unless he can recognize the different constellations—or at any rate identify them with the help of the maps that follow—he will not yet have begun to know his way about the night skies, nor will he be able to find the many objects of interest for binocular study (listed in Appendix 3) which each constellation contains. But before we can get on with learning the geography of the night sky we must first acquaint ourselves with the way in which it behaves as a whole.

The Diurnal Rotation of the Star Sphere

Quite a short period of observation will reveal the fact that the stars, like the sun and moon, rise in the east, climb up the sky until due south, and then sink again towards the western horizon. Furthermore, observation on two consecutive nights will prove that they make one complete circuit of the star sphere in about twenty-four hours—if a certain bright star is seen to be rising at eight o'clock one evening, it will be rising again at about eight o'clock the following evening. The stars are sometimes called the fixed stars, for the reason that although they move round the heavens in twenty-four hours they do not change their positions relative to one another in doing so; it is as though they were fixed immovably to the inner side of a great sphere, at whose centre is the earth, which rotates once every twenty-four hours, carrying the stars with it. Two stars no more change their distance apart or direction from one another during this rotation *en bloc* than do London and Edinburgh on a child's terrestrial globe change their relative positions when the globe is set spinning.

We have likened the star sphere to a child's globe, with the proviso that whereas we are looking at the latter from the outside we observe the former from the inside. The analogy between the two brings out one or two further points. Just as the child's globe rotates between two pivots, the north and south poles, and has a great circle (the equator) inscribed round it midway between these two poles, so does the star sphere rotate between two pivots—the north and south celestial poles—and so too has it an imaginary girdle known as the celestial equator. In England the north celestial pole is situated some 50° above the north point of the horizon, or a little more than half-way between the horizon and the point overhead; the south celestial pole is of course invisible, since diametrically opposite the north celestial pole—i.e. below the southern horizon somewhere under our feet.

The north celestial pole is easily found. Figure 1 illustrates the group of stars known as the Plough, part of the constellation of Ursa Major. Since it, in common with the rest of the star sphere, rotates about the north celestial pole, it may be found in any of the positions shown, or in some intermediate position. But it is nevertheless an unmistakable feature of the northern sky and easily identified. To give some idea of the size of the star group, the thick black line marked *Scale* represents approximately the size of a 6-inch rule held at arm's length against the sky. Having found this star group—somewhat resembling a saucepan with a bent handle —it is only necessary to carry one's eye away from the Plough in the direction indicated by the dotted line for a distance about equal to five times that separating the two stars marked β and α (known as the Pointers) for it to alight upon a star which, though rather faint, cannot be mistaken since no others lie near it. This star is Polaris, the Pole Star, and it marks almost exactly the position of

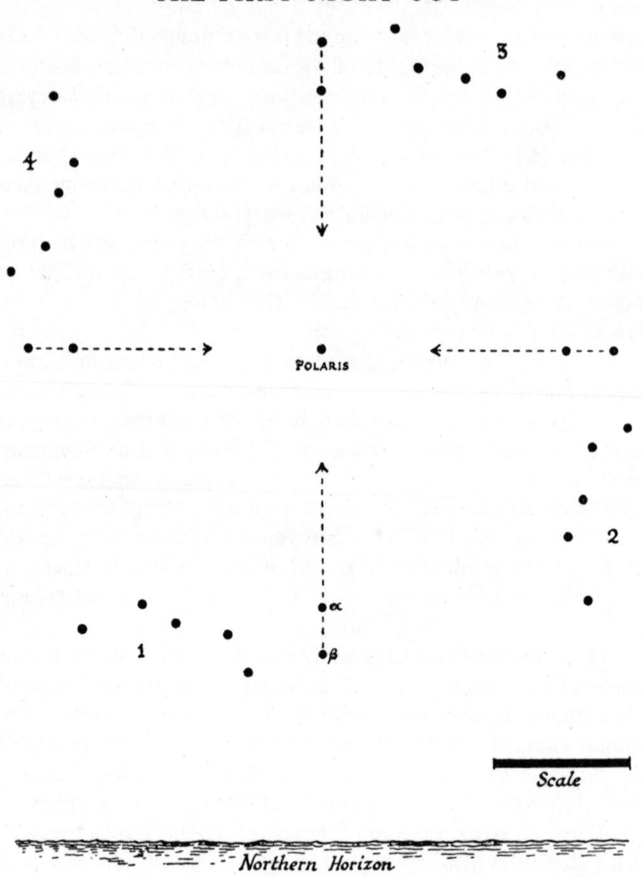

FIGURE 1

the north celestial pole. It can easily be seen from the diagram, or from the sky itself once the diurnal motions of the stars have become familiar, that any star whose distance from Polaris is less than the distance of Polaris above the northern horizon can never set: such stars, of which the Plough is an example, are known as circumpolar stars.

Let us summarize what we have discovered so far:

(i) The stars, like the sun and moon, rise in the east, set in the west, and make one complete circuit of the heavens in about twenty-four hours.

(ii) They carry out this motion *en bloc*, not altering their positions relative to one another.

(iii) The star sphere, a convenient fiction to which we may imagine the stars as being fixed, rotates about two diametrically opposite 'pivots', known as the north and south celestial poles.

(iv) Finally, we have located Polaris in the night sky and have seen that certain northern stars never set—the so-called circumpolar stars.

The Rotation of the Earth

The next point to get clear is why the star sphere behaves in this way. The ancients believed that the sphere was a material structure to which the stars were attached, like lamps hanging from a celestial ceiling, and that this sphere rotated in a period of twenty-four hours about a stationary earth. We now know, of course, that there is no such material sphere, that the stars are not all at the same distance from the earth, and that it is the earth's diurnal rotation from west to east which gives rise to the illusion of the stars revolving from east to west. It follows from this, as a little thought will show, that the celestial poles are the points where the earth's axis of rotation, produced in either direction, intersects the star sphere.

The Annual Rotation of the Star Sphere

But if the star sphere rotated in exactly twenty-four hours (to continue describing the phenomenon in these convenient but incorrect terms) the same stars would rise at the same time night after night and we should see the same constellations in the summer as we do in the winter. Now most people, ignorant as they may believe themselves to be of astronomy, know that this is not the case, and that the winter stars and constellations are not the same as those of summer. Why is this so? The answer is simply that the stars do not make one diurnal circuit of the star sphere in *exactly* twenty-four hours, but in a period some four minutes shorter. In other words the hands of the clock are moving more slowly than the stars—clock time is slower than star time. Let us see how this works out. A star situated on the celestial equator which is seen to be on the meridian[1] at 10 p.m. on one night will be on the meridian again at 9.56 the following night, 9.52 on the third night and 9.32 one week after the initial observation. At 10 p.m., therefore, it will now be well past the meridian. Three months later it will be setting below the western horizon at this hour and a new set of stars will be due south of the observer, stars that were just rising when the observations began. Thus it is that the compass bearings of the stars change from day to day, and each season has its own constellations (see the three-monthly maps, following p. 127). This slower rotation of the star sphere clearly occupies one year:

10 p.m. on day of first observation (say 1st Jan.): star on meridian,
10 p.m. on 1st April: star setting,
10 p.m. on 1st July: star invisible, since below the horizon directly below Polaris.
10 p.m. on 1st October: star rising above eastern horizon.
10 p.m. on 1st Jan.: star on the meridian again.

It is essential that the mechanism and nature of this seasonal change of the constellations (or *annual* rotation of the star sphere) be clearly understood before the actual identification of the different constellations is embarked upon. The reason for this rotation is that the earth is not only spinning on its axis (hence the diurnal rotation of the stars) but is also revolving about the sun in a period of one year, hence steadily changing the observer's viewpoint of the sun and stars from minute to minute and from day to day.

The Ecliptic, or Sun's Path

One other aspect of this annual rotation must be cleared up.

[1] The meridian is a great circle about the star sphere which passes through the north and south points of the horizon, the celestial poles and the zenith i.e. the point directly overhead).

How does the star sphere's yearly motion from east to west affect the sun? Does the sun partake of it? No, clearly not. For when we said that there is a daily four-minute discrepancy between star time and clock time we really meant that the discrepancy is between star time and solar time, for we set our clocks by the sun. A few days' observation will show that the sun is always on the meridian at midday (that, indeed, is the definition of midday), not four minutes earlier each succeeding day.[1] Since the star sphere is making its annual rotation behind the sun, the latter consequently appears to move across the background of the constellations from the west to east, completing one circuit in the same period of one year. The path that the centre of the sun's disc traces out upon the star sphere is known as the Ecliptic; this is clearly the line of intersection of the plane of the earth's orbit about the sun produced to meet the star sphere. And since the earth's equator is inclined to the plane of its orbit at an angle of 23½°, so the ecliptic is inclined to the celestial equator at this same angle.

The Zodiac

The Zodiac is an imaginary parallel-sided band, 18° wide, described round the star sphere, down the centre of which runs the ecliptic. To this narrow band are confined the sun, moon and planets. The zodiac is divided into twelve equal lengths, in each of which the sun is situated for one month of the year. These twelve constellations, to be described later, are called the Signs of the Zodiac.

The Motions of the Moon

We have now discussed and described the apparent motions of the stars and of the sun; those of the planets we will leave until Chapter 4, just remembering that the planets visible to the unaided eye never stray more than 9° from the ecliptic. There now remains the moon with its varying appearances. The motion of the moon against the background of the stars duplicates that of the sun, except that for one year is substituted one month. Thus the moon makes an apparent diurnal rotation of the star sphere in

[1] This is not always strictly true, but the discrepancies, on either side of midday, do not invalidate the argument.

about twenty-four hours, due to the earth's rotation. But owing to its own orbital motion about the earth, which is completed in twenty-seven days, it also makes a west to east circuit of the star sphere in that period. In other words, while being carried all the time from east to west along with the stars it has also a proper motion of its own from west to east. Thus it rises later each day: if it is on the meridian at 10 p.m. tonight, it will be considerably east of the meridian at 10 p.m. tomorrow night.

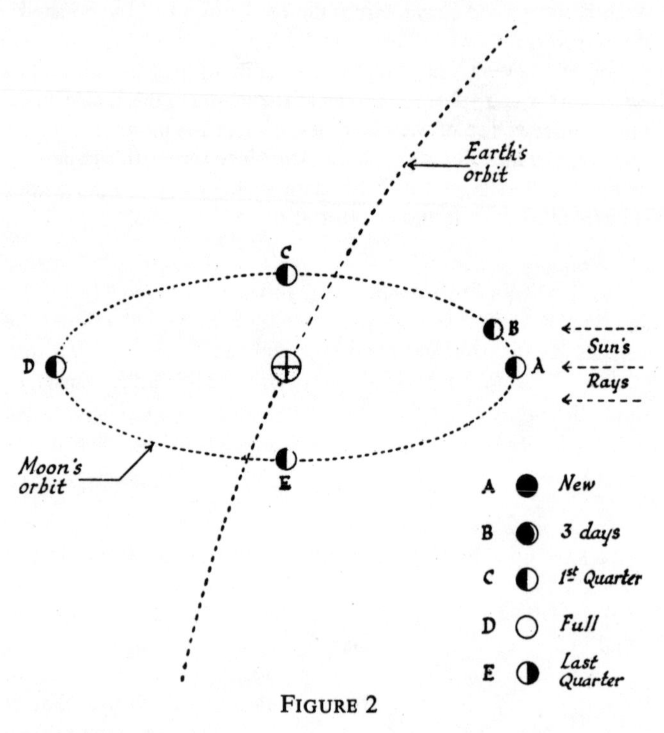

FIGURE 2

The Moon's Phases

So much for the moon's motions: what of its phases? Fig. 2 pictures the earth and moon, each revolving about its own orbit, and the direction from which the sun's light is falling. It is clear that only one hemisphere of the moon can be illuminated at a time, but the angle from which this hemisphere is seen by an observer on the earth varies with the position of the moon in its orbit. A moment's examination of this figure will show that the appearances of the moon in positions *A* to *E* are those shown at the side of the diagram: at *A* the illuminated hemisphere is turned directly away from the earth and the moon is therefore invisible (new moon); at *B*, a little later, we have a glimpse of the illuminated hemisphere; at *C*, we can see half of it, the right-hand or western half; at *D* the hemisphere is turned directly towards us and therefore the moon is fully illuminated (full moon); at *E* it has moved round to the other side of the sun and is now approaching it again—the half of the sunlit hemisphere that is turned towards the earth is the left-hand or eastern half; gradually during the last week of the lunation the illuminated portion shrinks to a narrow crescent, the moon being visible low in the eastern sky before sunrise; finally it becomes invisible altogether and it is new moon once more, four weeks after the preceding new moon.

Stars, Planets and Sun

Before proceeding with a detailed description of the Solar System it will be well to ensure that the reader is quite clear as to the difference between sun, moon, planets and stars.

The sun is a giant globe of incandescent gases and vaporised solids; in size and mass it compares with the planets as a motor bus compares with a matchbox.

The stars are essentially similar to the sun—in fact the sun *is* a star. That they look so different from the sun is simply a result of their much greater remoteness: even the nearest stars are 300,000 times as distant as the sun. For this reason the stars resemble the planets when seen with the naked eye—both are minute points of light. Yet when we reflect that this earth is a planet (and a fairly representative one) and the sun a star (again fairly typical), we shall see how different the two classes of body are: the difference between a star and a planet is the difference between the sun and the earth.

The moon may simply be regarded as a small planet. It is a cold, solid body in no way resembling the sun or the stars. It revolves about the earth and indirectly round the sun, being carried round the latter by the earth.

So far, then, we have discussed how the heavenly bodies move and behave. We have yet to learn something of what they are really like and then to start our exploration of the constellations. And if the reader feels that so far it has been rather stodgy going, let him reflect that no one has enjoyed the great literature of the world who has not undergone the preliminary tedium of learning his A B C.

Chapter 3

THE SUN

Its Status

The sun is nothing out of the ordinary in the universal scheme of things: merely one star among thousands of millions, and not even a particularly large or bright one; if removed to the same distance as Sirius, the brightest star in our northern skies, it would appear no brighter than Polaris. Nevertheless, the sun is the supreme ruler of the Solar System—that little family of planets and their moons, comets, meteors and asteroids. These all revolve around the sun, and are in complete subjection to its gravitational sway. The sun is the ultimate source of all forms of energy used on earth except nuclear energy, which is itself the source of the sun's own light and heat. It is also in a real, physical sense the earth's parent, and continues to affect its offspring in a number of unsuspected ways which will be described later.

Its Size and Distance

By terrestrial standards the sun is inconceivably remote; by astronomical standards, right on our doorstep—93 million miles away. Light, which travels so fast (186,000 miles in one second) that in everyday life we regard its passage as instantaneous, takes over 8 minutes to reach us from the sun; so that if the sun were by some divine *fiat* obliterated one midday, we should continue to see it shining in the sky until 12.08.

It is an incandescent sphere so vast that were it to replace the earth, the moon would be little more than half-way out to its surface. Its diameter is 864,000 miles, and one-third of a million earths could be made out of its material. The temperature at its surface is 6,000° C. (that of an electric welding arc is only about 3,500°C.), rising internally to a probable 20 million degrees at the centre. Its output of energy is so colossal that to express it in H.P. we should have to write a five followed by twenty-three noughts. Put in another way, it releases annually enough heat to melt a layer of ice 4000 miles thick at its surface.

The Analysis of Sunlight

So much of our knowledge of the sun has been given us by the spectroscope that it will be well, before embarking on a description of the appearance and nature of the sun, to give some account of the way in which this instrument analyses sunlight.

Light is a form of vibration, a system of waves. In this it resembles the radiation used in wireless, which differs from visible light only as regards its wavelength and frequency. The waves used in commercial medium wave radio broadcasting mostly measure between 10 and 1000 metres from crest to crest. Light waves, by comparison, are excessively minute: 1,250 waves of red light occupy a millimetre, and about twice that number of violet light.

Radiation whose wavelength is just too long to affect the human retina is experienced as heat, and is called infra-red. Much longer than the infra-red heat waves, but shorter than the radio waves, are those used in television and radar. Beyond the shortest waves of visible violet light lies the region known as the ultra-violet, and beyond that again the 'hardest' radiations of all—X-rays, the γ-rays released in radioactive decay, and cosmic radiation. (See Figure 3).

All these different wavebands together constitute the whole gamut of electromagnetic radiation; all are essentially the same, the only distinguishing feature among them being their wavelength.

Let us concentrate now on the visual waveband, that comprising violet, blue, green, yellow, orange and red light. When a solid body is heated to incandescence it emits radiation of every wavelength in this range. An incandescent gas or vapour, on the other hand, emits only a few selected wavelengths; each element radi-

The Range of Electromagnetic Radiation

(logarithmic scale)

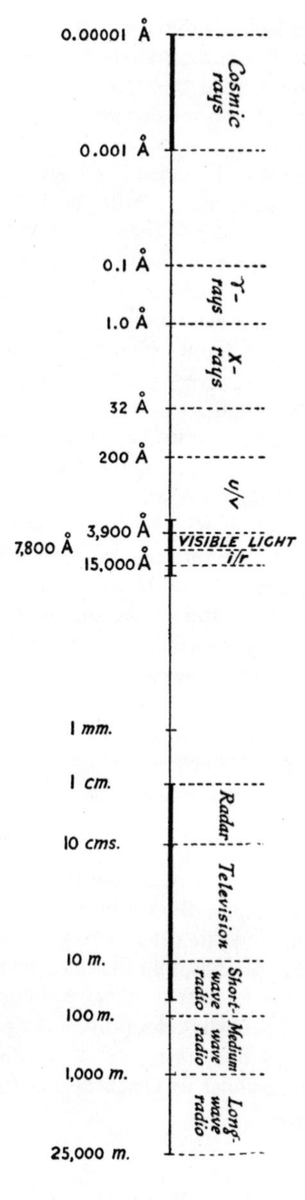

For the shorter wavelengths, the commonly used unit is the Ångstrom (Å)—one hundred-millionth of a centimetre. The wavelengths represented at the lower end of the scale exceed those at the upper by a factor of 25,000,000,000,000,000,000. The earth's atmosphere is opaque to the greater part of the range, 'windows' (the heavily printed sections of the scale) occurring at the extreme short-wave end, the visible section, and the short-wave end of the 'radio' section.

Figure 3

ates always the same wavelengths, which are different from those of every other element. If the polychromatic radiation from a glowing solid is made to shine through a gas or vapour which is cooler than itself, this gas will absorb those same wavelengths which it emits when itself radiating.

The construction of the spectroscope is described in Chapter 9, and here it need only be said that it sorts out, in order of wavelength, the light passing through it. Actually, it focuses the analysed light into a long strip, called a spectrum, with the longest wavelengths at one end and the shortest at the other (see Figure 14) The spectrum of a glowing solid will thus consist of a continuous strip of colour—violet at one end, merging into blue, green, yellow and orange, to red at the other end. Every point along this band consists of light of a single wavelength.

The spectrum of a gas will not be a continuous band of colour, since most of the wavelengths are missing, but a number of fine, bright, coloured lines. The colour of each line will depend upon its wavelength. Thus the spectrum of glowing sodium vapour consists of two yellow lines, and the reason for their being yellow will be obvious the moment we place a hotter source of polychromatic radiation behind the sodium vapour: the two bright sodium lines will instantaneously be converted into two dark lines crossing the yellow section of the continuous spectrum. Since the sodium is absorbing the light of these two wavelengths from the mixed radiation passing through it they will be missing from the latter's continuous spectrum, which will therefore be dark at the points corresponding to the two wavelengths.

The Sun's Spectrum

When sunlight is analysed in this way it is found that the spectrum consists of a continuous background, crossed by innumerable fine dark lines. We deduce, therefore, that the surface of the sun is emitting light of all visible wavelengths, and that overlying it is a cooler atmosphere which absorbs the light of the wavelengths corresponding with the lines. These Fraunhofer lines (named after their discoverer) are exceedingly numerous—nearly 25,000 have been catalogued from the ultra-violet to the infra-red. By measuring their wavelengths it has been possible to identify about sixty of the terrestrial elements in the sun's atmosphere.

Observing the Sun

Special precautions must be taken to protect the eye when observing the sun with even a small telescope or binoculars. Remembering that a simple magnifying glass can in a few seconds ignite a piece of paper on which the sun's image is focused, it will be understood that carelessness or ignorance in this matter can easily result in permanent blindness.

One of the simplest methods of protecting the eye is to cover the

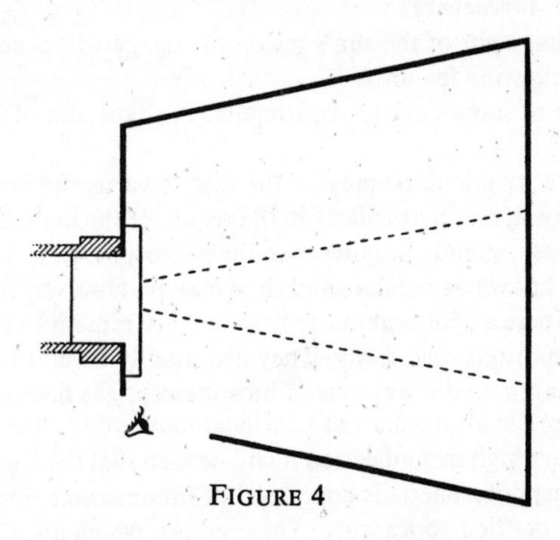

FIGURE 4

eyepiece with a sun-cap containing a piece of plane glass of a dense neutral tint. This is, however, a very unreliable method, unlikely to provide adequate protection for the observer. Another, and more satisfactory, for comfort, convenience, and safety, is to employ a solar diagonal—an eyepiece containing a piece of plain unsilvered glass set at an angle of 45° to the axis of the telescope. This unsilvered mirror allows most of the sun's heat and light to pass straight through it, reflecting only a small percentage to the observer's eye. Even so, a sun-cap (though a comparatively pale one) must be used as well.

But the safest and in many ways the best method of observing the sun is by projection. The sun's image is not viewed directly, but is thrown on a white screen. This can either be mounted or held behind the eyepiece in a darkened room, the telescope sticking out through a chink in the curtains; or else a light, strong box can be screwed on the drawtube by means of the eyepiece, the image on the back of the box being viewed through a hole in one side (Figure 4). It should be pointed out, however, that a terrestrial telescope (image the right way up) or any instrument with a 'field lens' may be damaged if used in this way.

Telescopic Appearance

One's first sight of the sun's telescopic image will probably reveal the following features:

(*a*) One or more dark spots, irregular in shape and of different sizes.

(*b*) A noticeable darkening of the disc towards the limb; near any large spot, and particularly in the region of the limb, fine hair-like markings, slightly brighter than the photosphere,[1] will be seen. These are known as faculae, and they may be observed to cluster together where a spot is about to form, and to remain so clustered after the spot has died away. They are usually several thousand miles in length, and are clouds of incandescent gas floating in the atmosphere far above the sun's surface; more about these later.

(*c*) Under high magnification it will be seen that the whole solar surface, or photosphere, is composed of innumerable tiny grains, giving it a mottled appearance. These grains, which are also more easily seen in the neighbourhood of spots, measure about 500 to 1,500 miles across, and thought to be the crests of great waves or upsurges of unequally heated photospheric material.

The Spots

Sunspots may be any size from a few hundred miles in diameter (the smallest that our telescopes will show) up to 100,000 miles or more. Indeed, a spot has to be about 25,000 miles in dia-

[1] The photosphere is the name given to the visible incandescent surface of the sun.

meter—three times that of the earth—before it can be seen with the naked eye.

From the particular way they are foreshortened when near the sun's limb, it is clear that spots are depressions in the solar surface —normally about 500 miles deep, but in exceptional cases nearly four times this figure.

A characteristic of the sunspots which the observer with a small instrument will quickly notice for himself is their tendency to occur in groups; especially in pairs, with or without a few smaller spots dotted around.

Telescopically, a large spot is seen to consist of a dark central area, the umbra, and a lighter fringe surrounding it, known as the penumbra. The fact that the spot is darker than the photosphere around it suggests that it is also cooler. This is confirmed by the spectroscope, which reveals in spot spectra the characteristic dark bands of chemical compounds. Now compounds are broken down into their constituent elements by heat, the temperature at which this happens depending on the particular compound. Titanium oxide, for example, is dissociated into titanium atoms and oxygen atoms at a temperature of about 3,000°; yet spot spectra commonly contain the bands of this compound.

Two other interesting facts concerning the spots are revealed by the spectroscope. They are centres of great cyclonic disturbances which reach for thousands of miles into the atmosphere above them. Apart from the strongly marked spiral configuration of the gases shown by the spectroheliograph in the vicinity of sunspots,[1] it appears that gases are being sucked into them at the higher levels and expelled from them in the vicinity of the photospheric surface.

Secondly, the spectroscope proves that each spot is the centre of a magnetic field, which in extreme cases may be almost a million times stronger than the earth's. By means of the so-called Zeeman effect—a broadening, doubling, or trebling of the lines of the spectrum—these magnetic fields have even been detected where no spot is visible, presumably indicating that the forces and

[1] See p. 43.

conditions which bring a spot into being are here present, but are not strong enough to have any visible result.

In the case of pairs of spots, it has been discovered that the two members are almost invariably of opposite polarity. Further, that if in the south hemisphere the eastern member of a pair is positive and the western negative, then in the north hemisphere it will be the western member which is positive and the eastern which is negative.

The Sun's Rotation

The length of life of the spots—like their size and shape—varies over a wide range. Generally speaking, the larger a spot is, the longer it will last. The smallest may live only a few days, while large ones may last for several months, and there are records of one spot which lasted for eighteen months, though this is most unusual.

Observation of the same spots from day to day will reveal that they are drifting slowly across the sun's disc from east to west, owing to the sun's rotation. When allowance is made for their own individual motions—which are comparatively small—it is possible to deduce the period of the sun's rotation from their observation. A very curious fact then emerges: the sun does not rotate as a solid body at all, but at different speeds in different latitudes. At the equator one rotation is accomplished in about 24½ days; in latitude 45° N/S the period has lengthened to 27½ days; while in the vicinity of the poles it is about 33½ days.

The Spot Cycle and the Law of Zones

During the early years of last century a German amateur astronomer, Schwabe by name, observed the sun with his small telescope and daily counted the spots that were visible. In 1843 he reached the epoch-making conclusion of his twenty-seven years' work: the number of spots on the sun is not even approximately constant, but varies from a definite minimum up to an equally well defined maximum and back to minimum again in a period of about eleven years.

This sunspot cycle, which appears to play a fundamental role in the sun's internal economy, applies to the number of spots and not to their size—large and small spots occur indifferently throughout the cycle. At spot maximum it is no uncommon thing for twenty-five or thirty spots to be visible at the one time, and the sun is rarely quite spotless. At minimum, on the other hand, it may be free from spots for weeks at a time.

A new eleven-year cycle began early in 1944, and within two years had produced the largest spot since the Greenwich records began in 1874. Five months later another giant spot, the second largest ever recorded, made its appearance. Both these, however, were surpassed by the spot of April 1947, which involved about

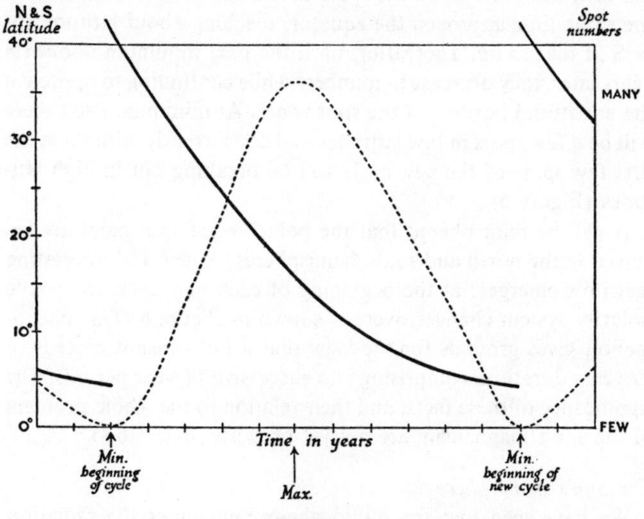

FIGURE 5

Diagram showing both the numbers and the distribution of sunspots throughout one 11-year cycle. The scale at the bottom of the figure represents years. The vertical scales represent spot numbers (dotted curve) and the latitude in both hemispheres at which they occur (continuous curve). Thus, from Minimum the numbers increase to a Maximum about 5 years later, and then fall off again to the next Minimum. The latitude curves show that at Minimum there are a few spots from the preceding cycle in low latitudes as well as the earliest ones of the new cycle in high latitudes. As the cycle progresses, the latter approach the equatorial region.

5,500 millionths of the sun's hemisphere, or over 6,000 million square miles.

Spots do not occur all over the sun, but are confined to two zones, one in each hemisphere stretching roughly from latitudes 5° up to 35° or 40°. Few spots occur within about 5° of the equator, and virtually none between 40° and the poles.

However, they are not to be found scattered all over these two zones at all times, their distribution in latitude being intimately connected with the spot cycle, as Spörer first pointed out. At the beginning of a cycle (spot minimum) a few spots break out at the polar extremities of the two zones, about 35° to 40° N/S. During the next five years or so the spots become more numerous and at the same time approach the equator, reaching about latitude 15° N/S at maximum. Thereafter, until the next minimum about six years later, they decrease in numbers while continuing to approach the equatorial borders of the spot zones. At minimum itself there will be a few spots in low latitudes and concurrently with them the first few spots of the new cycle will be breaking out in high latitudes (Figure 5).

It will be remembered that the polarities of spot pairs are reversed in the north and south hemispheres. A new and interesting fact now emerges: at the beginning of each new cycle the whole polarity system changes over, as shown in Figure 6. This phenomenon gives grounds for the view that a full sunspot cycle is of 22 years duration, comprising two successive 11-year periods. The significance of these facts, and their relation to the whole problem of the sun's magnetism, are not yet properly understood.

The Sun's Atmosphere

We have seen that the photosphere contributes the radiation of all wavelengths which produces the continuous background of the solar spectrum; and that the dark monochromatic lines which cross it indicate that absorption of certain wavelengths is occurring in a cooler atmosphere overlying the photosphere.

As the moon passes across the sun's face in the early stages of a total solar eclipse, a moment arrives when the photosphere is entirely hidden but a narrow crescent of the sun's atmosphere is still

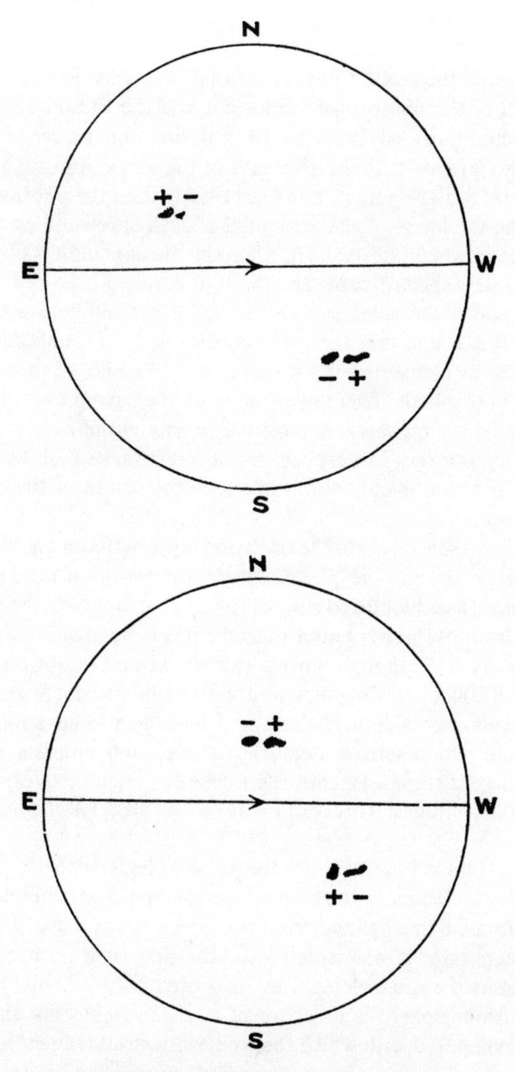

FIGURE 6

Upper: Last period, from Minimum 1944 to Minimum 1954.
Lower: Current period, from Minimum 1954 to Minimum ?1965.

visible round the moon's limb. When this happens the continuous spectrum of the photosphere fades out and the Fraunhofer spectrum is simultaneously reversed, i.e. it flashes out as a set of *bright* lines. This proves that the elements in the sun's atmosphere are themselves radiating light, but less brightly than the photosphere From the thickness of the lines of the 'flash spectrum' and from the order in which they vanish, when the phenomenon is observed with a slitless spectroscope (see ch. 9), it is possible to deduce the heights above the solar surface to which the different elements extend. It is found that the vast majority of the Fraunhofer lines are caused by elements lying within about 500 miles of the surface of the photosphere. This lowest layer of the sun's atmosphere is called the reversing layer. It must not be thought of as in any way resembling our own atmosphere for, apart from its high temperature, its pressure is only about one ten-thousandth of the earth's at sea level.

Merging insensibly into the reversing layer, without any definite boundary, is the chromosphere. During the period of total eclipse this appears as a blood-red rim completely surrounding the moon.

The chromosphere is hotter than the reversing layer, and about ten times as deep, though ionized calcium occurs exceptionally as high as 8,000 or 9,000 miles above the sun's surface. Its chief constituents are calcium, helium and hydrogen (whose red light gives it its characteristic colour). Its spectrum consists of the bright lines of these elements, the numerous metallic lines, which are so conspicuous a feature of the reversing layer spectrum, being absent.

At the total eclipse of 1868 the yellow chromospheric line of an unknown element was observed for the first time; this element was given the name helium (from the Greek *helios*: sun). Twenty-seven years later it was isolated for the first time on earth and identified as the same element by its spectrum.

The chromosphere is a region of continuous storms and upheavals compared with which the worst terrestrial tempest is but a gently wafting zephyr. Great jets and prominences of material are ejected from it at velocities which often approach a million miles an hour, and to distances from the sun's surface which may exceed half a million miles. An astonishing feature of these eruptive prominences is that they frequently accelerate as they shoot outward from the sun—one would expect them to decelerate and finally fall back into the sun. Not only that, but they accelerate in jerks. The motions of the eruptive prominences are still very inadequately understood, as is the whole subject of the physical conditions of the solar atmosphere.

In addition to the spectacular eruptive prominences there is a second type, known as quiescent prominences. These are much more stable than the former, and appear as great projections pushed up from the upper regions of the chromosphere. Their average height is of the order of 50,000 miles, and they are seldom less than 50,000 miles long, exceptionally ten times this figure.

Prominences, unlike the spots, occur in all latitudes, though they are most numerous in the spot zones. Their numbers, however, as well as their general distribution, vary with the eleven-year cycle.

Until 1868 the prominences could only be observed during the short and infrequent periods of total eclipse. In that year a great advance was made, with the discovery of a method whereby they could be observed at any time. If the slit of a spectroscope is laid tangentially to the sun's limb at a point where there is a prominence, the bright lines of the latter's spectrum will be observed. Now the effect of using a spectroscope of high dispersion is to weaken the intensity of the continuous background (since spreading it over a greater area) while at the same time leaving unaffected the intensity of the bright lines, which are caused by radiation of a single wavelength. If the slit of a high-dispersion spectroscope is adjusted as described above, and then opened slightly, each of the bright lines will widen into a monochromatic image of the prominence.

The Corona

The outermost region of the sun's atmosphere, the beautiful corona, was likewise visible only during total eclipse until the ingenuity of the astronomer devised a method of seeing it in full sunlight. The scale of this achievement will be appreciated when it is learnt that the total brightness of the corona does not exceed about half that of the full moon.

At total eclipses the corona appears as a pearly white radiance surrounding the black body of the moon. It may be of more or less equal depth all round (nowhere less than about 300,000 miles), or it may be drawn out into long petals and rays up to 5 million miles long. Its general shape varies in step with the eleven-year cycle.

A rough distinction between an inner and an outer region of the corona can be made, based chiefly on spectroscopic differences. The inner corona is much the brighter, and its spectrum consists of some thirty bright lines superimposed upon the continuous photospheric background. Some of these lines, discovered at the eclipse of 1898 had, and still have, no counterpart in terrestrial laboratory spectra; they were accordingly attributed to an unknown element, 'coronium'. But the story of helium was not to be repeated: their origin remained a mystery until 1941 when Edlén demonstrated that they are not caused by an unknown element, but by well-known substances—chiefly iron—behaving in a manner only possible under the peculiar physical conditions (high temperature and abysmally low pressure) which obtain in the corona. It is quite impossible to reproduce these conditions in the laboratory, and therefore the mysterious lines cannot be observed in any experimental spectrum.

The spectrum of the outer corona, which is much fainter, is a faint replica of the Fraunhofer spectrum superimposed on a continuous background. For this reason it is thought that it shines mainly by reflected sunlight.

The problem of observing the corona without an eclipse is the mechanical and optical one of reducing the diffused light of the photosphere sufficiently to enable the incomparably less brilliant corona to be seen. This scattering of sunlight has two main sources: in the earth's atmosphere, where it is scattered by droplets of water vapour and particles of soot and dust, as well as by the atmospheric molecules themselves; and in the telescope, from the surfaces of each lens and from bubbles and other flaws in the glass.

In 1931 the French astronomer Lyot succeeded in building a telescope of special design and materials which eliminated the greater part of this second source of diffusion. By setting up his instrument at the observatory on the summit of the Pic du Midi he was able sufficiently to reduce the first. He was successful in photographing the spectrum of the corona, and later the corona itself, in full sunlight.

The Sun in Monochromatic Light

If a certain area of the sun is emitting light of a single wavelength it will not be visible in the telescope, since it will be swamped in the flood of polychromatic radiation coming from the photosphere. But if we could eliminate this light of other wavelengths from the infra-red to the ultra-violet, leaving only the one wavelength in question, then we should be well on the way to seeing the object that is radiating it.

This is what is accomplished by the spectroheliograph, one of the most ingenious pieces of apparatus that astronomers have devised to sharpen their perception of the hidden aspects of the universe.[1]

Selecting, say, one of the bright lines of calcium and excluding radiation of every other wavelength, we obtain a picture of the sun which shows the incandescent calcium and nothing else. The distribution of solar calcium and hydrogen is particularly interesting because it is not uniform: these elements are encountered in the solar atmosphere in great clouds, or flocculi.

Calcium flocculi may be either bright or dark, according to their height in the atmosphere, whence their temperature. They are characteristically rounded and compact in form, and although they are found all over the disc they congregate particularly over sunspots. Hydrogen flocculi, on the other hand, are nearly always dark, being at higher atmospheric levels than the bright calcium clouds. They are typically wispy and filamentous in shape. Like the calcium flocculi, they tend to gather over disturbed areas of the sun's surface, such as spot groups, and it was their radial, Catherine-wheel-like arrangement over large spots (shared to a

[1] The instrument is described in Chapter 9.

smaller extent by the calcium flocculi) which first suggested that the spots are cyclonic vortices.

When seen near the limb, hydrogen flocculi are often found to be associated with prominences, and it is certain—at any rate in some cases, and perhaps in all—that what appears as a prominence when seen in profile on the sun's limb is a hydrogen flocculus when seen projected on the disc. Faculae, in the same way, are associated with, or identical with, bright flocculi.

Like so much else on the sun, the flocculi wax and wane in number with the eleven-year cycle.

A fairly recent discovery of the spectroheliograph is the solar 'flare'. What precisely a flare is, or what causes it, is not yet known; it appears to be a sudden outburst or eruption, always associated with an active spot, from which is emitted a flood of hydrogen or calcium light, as well as ultra-violet radiation and electrified particles. Flares are always short-lived, some lasting only a few minutes and few longer than an hour. They are of particular interest for their effects upon human activities on this planet.

The Sun and the Earth

As our knowledge of the sun has increased, so has more and more been learnt of the various and often unexpected ways in which solar events affect conditions on earth.

Many attempts have been made to correlate terrestrial events with the eleven-year cycle—unsuccessfully in the case of specifically human concerns such as trade booms and depressions, epidemics, and the rate of growth of children.

More successful has been the search for a reflection of the eleven-year periodicity in the earth's weather, and in fields indirectly determined by the weather. Douglass, for example, has found a periodicity of this length in the growth rate of trees (determined by measuring the thickness of the annual growth-rings) and has carried his researches back over a period of several thousands of years. The levels of lakes and the size of certain animal populations (fish, for economic reasons, have been specially studied) exhibit a similar periodicity, though the correspondence is not infallible. Even the number of rabbit skins brought in to the Hudson

Bay Company has over a period of half a century fluctuated apparently in step with the solar cycle!

Short-term weather correlations often appear to work for a time, only to break down completely: even if such correspondences are not coincidental there are clearly many other factors involved. But in general it seems that the fall of both rain and snow tends to be heavier at spot maxima than at minima; the barometric pressure, the mean annual temperature, and the incidence of hurricanes may also be related to the solar cycle, but the evidence is far from conclusive.

When we come to those terrestrial events that are directly dependent upon sunspots and flares, we are on much firmer ground. Not only is the eleven-year periodicity clearly marked, but in most cases it is possible to recognize the operation of cause and effect between a specific solar event and its terrestrial result.

The earth is a magnet to whose poles the compass needle points in the two hemispheres. Small daily variations in its magnetic field occur all the time, and these diurnal variations are more marked at spot maximum than minimum. Sudden and great variations in the field are called magnetic storms, and these too are much more common around maximum. The arrival of a large spot at the central meridian of the sun's disc frequently heralds a magnetic storm; still more certain is it that a flare will be followed by marked magnetic disturbances. The onset of the magnetic storm lags behind the solar event by a variable interval which averages about twenty-five hours. The agent is therefore not radiation (which would reach us simultaneously with the visual observation) but streams of electrons emitted from the disturbed area of the photosphere, which travel at a mean velocity of some 1,100 m.p.s.

That these particles are not shot out in all directions from the spot, but in a comparatively narrow beam, is indicated by the facts that a storm follows the spot's crossing of the central meridian (when it is facing the earth), that the storms are of short duration, and that some of the most violent ones are repeated after an interval of twenty-seven days (when the sun's rotation will have carried the disturbed region back to the same position relative to the earth).

After a lag of the same length—and presumably therefore also caused by electron beams expelled from the sun—heavy earth currents, induced currents in telephone and telegraph cables, and aurorae may occur.

When an electric current is passed through a gas under low pressure—as in a neon advertising tube—the gas glows. In the upper atmosphere the requisite low pressures are encountered, and the part of the electric current in the discharge tube is played by the streams of electrified solar particles: these may ionize the gases of the atmosphere so heavily that a high enough potential is created to produce a visible discharge. This glow in the upper atmosphere (commonly several hundred miles up, and exceptionally as high as 700 miles) is called an aurora.[1]

Interference to short-wave radio reception, on the other hand, is synchronous with the visual solar event: there is no 24-hour lag. The agent in this case is ultra-violet light, which appears to be emitted in a flood by flares, and which by altering the electrical properties of the ionized layers in the upper atmosphere (120 to 170 miles high) from which the short waves are normally reflected, distorts or completely eliminates long-distance reception. This occurs only on the sunlit hemisphere of the earth; on the night side, which is shadowed from the ultra-violet radiation by the body of the earth, no interference occurs.

The intensity of the solar ultra-violet varies by at least 100 per cent between spot maxima and minima, whence short-wave radio interference is markedly more common at the former times.

The Sun's Radiation

Visible light, ultra-violet and infra-red do not make up the sum of the sun's radiation. One of the most surprising developments of the past few years was the discovery that the sun is also emitting the much longer radiation to which wireless receivers are sensitive: in other words, it is broadcasting.

[1] In August 1947 radar echoes were obtained from an aurora for the first time; its height was about 320 miles.

The strange story of the 'solar noise' or 'solar static' began one day in February 1942, when military radar sets in scattered parts of England apparently went haywire: they seemed to be giving echoes from a perfectly empty sky, innocent of enemy as well as of allied aircraft. During the night the interference stopped, but the same thing happened the next day. When the reports were correlated it was realized that in every case the direction from which the 'echo' came was within a few degrees of the sun. The sun, in fact, was broadcasting on the radar wavelength. It is significant than an eight-hour radio fadeout and an intense magnetic storm occurred during the same period, and that a large and active spot crossed the sun's central meridian and several flares were observed during the days in question.

Research since that date has shown that radio-frequency radiation is being emitted by the sun all the time. In a radio receiver tuned to certain wavelengths between 1 and 30 metres, it sounds like a steady hiss. The strength of the static varies roughly with the spottedness of the sun's disc, stronger 'bursts' being caused both by large spots and by flares.

Solar Eclipses

In its monthly journey round the earth, the moon sometimes passes directly between it and the sun. That it does not do so every month is due to the fact that the moon's orbit is slightly inclined to that of the earth, which it intersects at two opposite points. Usually, therefore, it passes above or below the sun at new moon, an eclipse occurring only when the line from the earth to the sun, and from the earth to the points where the lunar and terrestrial orbits intersect, are close together.

In each year there must be two solar eclipses, and may be as many as five. An eclipse may be partial; or the moon may pass centrally across the sun's disc, when the eclipse will be either total or annular. At a total eclipse, the sun is completely hidden by the moon; at an annular eclipse, a narrow rim of the photosphere remains visible round the moon's limb. Whether a central eclipse is total or annular depends upon the moon's distance from the earth at the time. The length of the moon's shadow averages

232,000 miles, but its distance from the earth may be anything from 222,000 to 253,000 miles. When farthest from the earth, therefore, its shadow does not quite reach us: it will appear slightly smaller than the sun, and the eclipse will be annular.

The strip of the earth's surface which is traced out by the core of the moon's shadow, from which the sun is seen to be totally eclipsed, cannot exceed 168 miles in width. Totality will then last for 7½ minutes, but usually it is very much shorter than this, a totality of even 5 minutes being rare. An annular eclipse, on the other hand, may be seen as such from a belt of the earth's surface as wide as 230 miles, and a partial eclipse from one to about 4,000 miles wide.

Although total eclipses are by no means rare, they recur at a given position on the earth's surface only at very long intervals indeed. Before the invention of the open-slit and coronograph techniques they provided our only opportunities for studying the prominences and corona. Nowadays they are less scientifically vital, but are nevertheless awe-inspiring spectacles, and no one lucky enough to have witnessed a total solar eclipse will forget the experience.

Chapter 4

THE PLANETS

The Structure of the Solar System

The planets are nine in number, each a world of the same general type or status as the earth. They all revolve about the sun in roughly circular orbits, the length of time taken to complete one such revolution (i.e. the planet's year) depending upon its distance from the sun. All these nine orbits lie in nearly the same plane,[1] and thus it is that no planet can stray more than a few degrees from the ecliptic. If you visualize the Solar System situated at the centre of the star sphere and yourself as situated on the earth in the general plane of the orbits, you will appreciate that these other orbits, as silhouetted against the starry background, will all lie very close to one another.

The names of the planets in order from the sun outward are: Mercury, Venus, Earth, Mars, Jupiter, Saturn, Uranus, Neptune, Pluto. Thus Mercury and Venus are always nearer the sun than the earth is, and the other planets farther from it. The following figures are condensed from the table on p. 129:

Planet	*Distance from Sun (in million of miles)*	*Period of revolution round the Sun*
Mercury	36	88 days
Venus	67	225
Earth	93	365
Mars	142	687
Jupiter	483	11·9 years
Saturn	886	29·5
Uranus	1,782	84·0
Neptune	2,793	164·8
Pluto	3,670	247·7

[1] With one exception, see p. 58.

The Solar System to Scale

Such distances are so much greater than those with which we are familiar in our everyday lives that they are difficult to visualize. To help overcome this difficulty let us construct a scale model of the System. For the sun, whose diameter is 864,000 miles, we will have a globe 150 feet in diameter; we will suppose that it is situated in Trafalgar Square, where it will completely engulf Nelson's Column.

On this scale Mercury will now be a 6-inch ball revolving about Nelson's Column at a distance of just over a mile—past Victoria Station and Holborn Viaduct.

Venus, 16 inches in diameter, will pass near the Tower of London, the Zoo, and Paddington, in an orbit whose radius is 2¼ miles.

The earth lies still further out, at a distance of 3 miles from Trafalgar Square. Its orbit will carry it through Notting Hill Gate, Whitechapel and Clapham Junction. Like Venus, it will be represented by a 16-inch ball. Forty feet away, and revolving round it, will be a 4½-inch ball, the moon.

The 9-inch ball representing Mars will pass through Hammersmith, Greenwich and Golders Green, at a distance of nearly 5 miles from Trafalgar Square.

Jupiter, the giant among the planets, with a scale diameter of over 15 feet, will be found 16 miles away: Epsom, Watford and Epping lie on its orbit.

Saturn, only slightly smaller than Jupiter (diameter, 12 feet) will take in Hitchin, Guildford and Tunbridge Wells.

Uranus will be represented by a globe 5½ feet in diameter, passing through Chichester, Reading and Clacton, 60 miles from London.

Neptune, a ball 4 inches larger than Uranus, lies 93 miles out: Cheltenham, Leicester and Calais.

Finally comes Pluto, a ball about 8 inches in diameter, whose orbit 122 miles in radius will take it over English soil during only half of its revolution; it will pass near Skegness, Hereford and Yeovil, across the Channel to Abbeville, and thence out across the North Sea to skirt the Norfolk coast.

The Apparent Motions of the Planets

How do these circum-solar motions appear to us, situated on the earth and observing the other planets against the background of the stars? Let us take the inner planets, Mercury and Venus, first. Fig. 7 shows the sun, the earth, and an inner planet in four positions along its orbit; for simplicity's sake we will suppose that the earth does not move appreciably during the time required for the planet to complete one revolution about the sun—the truth of the account is not materially affected by so doing. When the planet is in position *A* it occupies the same region of the zodiac as the sun, i.e. is situated in the sky during the day and below the horizon at night, and is therefore invisible. As it moves towards *B* it will re-

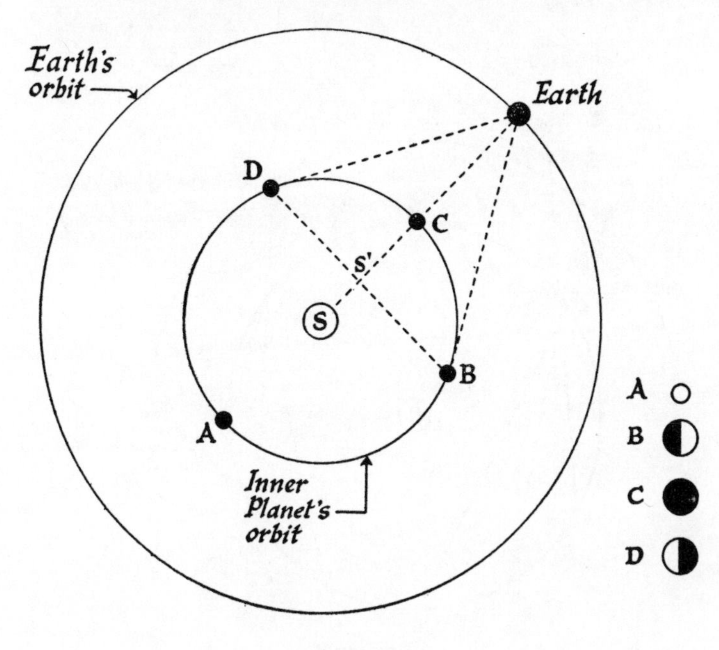

FIGURE 7

cede further and further from the sun until it is as far to the left or east, of the sun as it is possible for it to go: it is an evening star. From *B* to *C* it once more approaches the sun and becomes invisible, only to reappear on the other (right-hand, or west) side of the sun and recede from it until it reaches maximum elongation west at *D*. Once again the angular distance of the planet from the

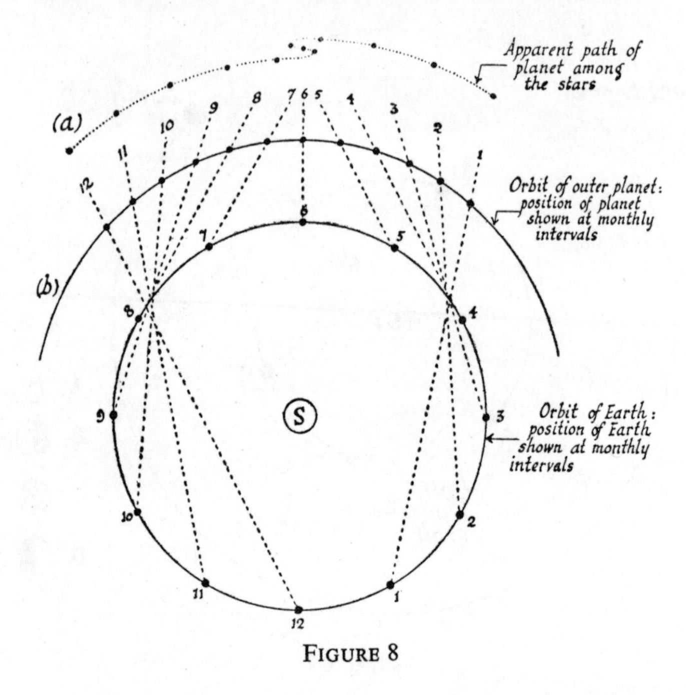

FIGURE 8

sun decreases, till position *A* is reached. An inner planet, therefore, can never recede beyond a certain fixed distance along the zodiac from the sun; it appears alternately as a morning star (visible above the eastern horizon before sunrise) and as an evening star (visible above the western horizon after sunset). The maximum elongation (the angular distance *S′D* or *S′B*) of Mercury, the nearest planet to the sun, is 29°; of Venus, with an orbit nearly twice as large as Mercury's, some 47°. Furthermore, in travelling round the orbit it will exhibit the same cycle of phases as the moon, as shown at the edge of the diagram.

But the outer planets behave quite differently. Since the earth completes the circuit of its orbit in a considerably shorter period than these planets (see Table, p. 119), the effect of the terrestrial observer 'catching up' and 'passing' the outer planets has to be added to the effect of their own west-east motion among the stars, as is demonstrated by Fig. 8 (*b*). It will be seen that the combined effect of the outer planet's own orbital motion and the orbital motion of the terrestrial observer is to make the planet appear to move across the starry background in a series of loops (Fig. 8 (*a*)).

The stage is now set for brief descriptions of the planets themselves.

Mercury

The nearest planet to the sun (36 million miles), and the smallest, its diameter of about 3,000 miles being less than half that of the earth. In many respects Mercury resembles our moon; the size and mass of the two bodies are very similar, for instance, and like the moon it has no atmosphere. Only the vaguest markings have been detected on the Mercurial surface—even these require large telescopes to be seen at all clearly. Its period of rotation about its axis has only recently been established as 59·4 days.

Venus

Revolves round the sun at a distance of about 67 million miles in a period of 244 days. If Mercury resembles the moon, Venus is almost the twin sister of the earth:

	Earth	*Venus*
Diameter	7,927 miles	7,600 miles
Volume	1·00	0·88
Mass	1·00	0·83

The chief cause of Venus's surpassing brilliance is the fact that it is sheathed in a dense atmosphere of clouds, the solid surface of the planet being entirely hidden from our eyes. White clouds reflect, after snow, more of the light falling on them than almost any other substance. The chemical constitution of this deep and dense atmosphere is still in doubt, though carbon dioxide has been identified spectroscopically. Water vapour and oxygen do not appear to be present. The mystery of the surface of this planet will be solved only when many interplanetary probes have penetrated the opacity of its atmosphere.

Mars

The next planet to the earth on the side farther from the sun. Its distance from the sun is 142 million miles, and when nearest the earth only 35 million miles separate the two planets; at such times Mars is brighter than any star and is surpassed only by Venus and by Jupiter when at its brightest. It is the only planet on which surface markings can be observed and studied in any detail; for it is comparatively near the earth, has a clear atmosphere and, being an outer planet, can be observed on the meridian (the most favourable position) during the night.

Mars is a smaller body than the earth (respective diameters, 4,200 and 7,927 miles), but in many respects the physical conditions and its surface resemble those of the earth more closely than any other planet. The spectroscope shows that the Martian atmosphere contains water vapour (clouds are often visible telescopically, floating above the Martian surface), oxygen, though in smaller quantities than the terrestrial atmosphere, and carbon dioxide; no toxic gases have been detected or suspected. Its temperature at the point where the sun is directly overhead varies from about freezing point to some 25° C. Elsewhere on the sunlit hemisphere it will of course be colder than this, and on the night hemisphere very cold indeed by terrestrial standards. It rotates on its axis in 24 hrs. 37⅓ mins.

The surface markings consist of blue-green areas superimposed upon the general reddish background of the planet. In addition to these there are a number of fine streaks known as *canali*. There is no evidence that these are artificial canals or even that they are waterways. Finally, there are two 'polar caps'. These behave in the same manner as the terrestrial polar caps, shrinking in summer and increasing in size during the winter. They were accordingly thought to be composed of snow, ice, or hoarfrost, but more recent work has suggested that they may be atmospheric phenomena—great patches of fog, mist, concentrations of ice crystals in the air, or whatnot.

The dark markings undergo a regular seasonal change, associated with the 'melting' of the polar cap in the hemisphere which is turning from winter towards summer. It has been suggested that this annual darkening of the blue-green areas in the spring hemisphere may be the proliferation of some form of vegetation under the stimulus of increasing temperature and the liberation of moisture from the polar cap. The darkening could, however, equally well be explained in terms of the tendency of many substances to darken somewhat when damp. All we can safely say on the vexed question of life on Mars is: *if* there is life resembling the terrestrial form in its fundamental features anywhere in the Solar System besides this planet, then Mars (with its near-terrestrial conditions) is the place where we might most reasonably hope to find it.

Mars has two satellites, discovered in 1877, and only visible in large telescopes. They are small—possibly only about 5 and 10 miles in diameter—and their behaviour is so curious as to give them a unique position among the thirty known satellites in the Solar System. Phobos is so near to the Martian surface (less than 4,000 miles) that it would never be visible above the horizon to an observer situated within 20° of either of the Martian poles. It completes one revolution round the planet in 7½ hours: in other words, there are three Martian months, as determined by this moon, to every Martian day! The revolution of Deimos, 12,500 miles from the planet's surface, is only 6 hours longer than the Martian day, with the result that this satellite remains above the horizon for 3 days at a stretch, passing twice through its complete cycle of phases during that time.

Jupiter

The giant planet of the Solar System. Its equatorial diameter of 88,700 miles is eleven times that of the earth, while its volume and mass are respectively 1,312 and 318 times as great as the earth's. Jupiter therefore contains enough material for a complete duplicate set of the remaining eight planets.

This great body has a retinue of twelve satellites, four of which were discovered by Galileo—one of the first discoveries of the newly invented telescope—and may be seen with binoculars: the diameter of two of them exceeds 3,000 miles (that of our moon is 2,160) while the other two are more than 2,000 miles in diameter. The remaining eight satellites are small and faint, the tenth and eleventh only being discovered in 1938, and the twelfth in 1951.

When viewed with a telescope, even the smallest, Jupiter is seen to be crossed by a complex system of dark belts. The rapid changes that occur in the detail of these prove them to be atmospheric: here, as in the case of Venus, the surface of the planet is hidden from our probing eyes by a dense vaporous atmosphere. The spectroscope shows that two of the main constituents of this atmosphere are methane (marsh gas) and ammonia. When we reflect that the temperature of the planet is some 130° C. below freezing point we can appreciate that Jupiter is no hospitable home for life as we know it.

Irregularities in the belts, white and dark spots, and other details enable the astronomer to measure the rotation period of Jupiter. The results of such measurements were surprising, and confirmed the atmospheric nature of the Jovian surface which we see. For it transpired that Jupiter, like the sun, is rotating at different speeds in different latitudes. The rotation period in equatorial regions is 9 hrs. 50 min., while towards the poles it is 5 min. longer.

Saturn

Unique as being the planet with the rings, Saturn is a giant planet only slightly smaller than Jupiter. Like its greater neighbour it is swathed in clouds, its surface being entirely hidden from us. It has a family of ten satellites, a few of which are visible in small telescopes and the largest of which should be glimpsed with good binoculars; this satellite, Titan, is bigger than Mercury. Saturn's temperature must be even lower than that of Jupiter, for its distance from the sun is nearly twice as great.

The rings lie in the equatorial plane of the planet and are composed of a vast number of separate fragments. The inner diameter of this vast structure is about 93,000 miles, its outer diameter 170,000 miles. Its thickness is small compared with its width—probably less than 10 miles—and when the rings are turned edge on to the terrestrial observer they vanish from sight.

Uranus and Neptune

Both are smaller planets than the giants Jupiter and Saturn, their diameters being about 32,000 miles, or four times that of the earth. Yet owing to their enormous distances both from the sun and from the terrestrial observer they are always faint features of the night sky and eluded detection till 1781, when Herschel discovered Uranus, and 1846 when Galle discovered Neptune. Nothing is known of their physical conditions, but their temperatures must be very low indeed, and the spectroscope proves that they have cloud-laden atmospheres like Jupiter and Saturn. Uranus has a family of five satellites, Neptune two. Uranus V was discovered photographically with the 82-inch reflector of the McDonald Observatory in 1948, Neptune II with the same instrument in May 1949.

Pluto

Pluto was discovered as recently as 1930. It is only visible in the largest telescopes, and was discovered photographically. Its mean distance from the sun is 3,670 million miles (nearly forty times greater than the earth's), at which distance it exhibits a sensible disc in none but the largest telescope; at this distance, too, it receives only 1/1,600 as much solar heat and light as the earth does—its temperature is probably in the region of –250°. To an observer on Pluto the sun would appear as a disc the same size as that of Jupiter seen from the earth, but some 650 times (7 mag.) brighter than the full moon. It is smaller than Mars, its diameter probably about 3,750 miles. It completes one revolution round the sun in 248 years and its orbit has the extraordinarily high inclination to the ecliptic of 17°. Pluto, therefore, is the one planet that can stray beyond the limits of the zodiac.

It is thought unlikely that Pluto has a satellite, though even if it had, existing equipment would probably be incapable of showing it.

The Planets in the Night Sky

Of the nine planets which are known to circle round the sun, two, Neptune and Pluto, can never be seen with the naked eye; a third, Uranus, is so faint that although just visible without instrumental aid it does not 'catch the eye', being indistinguishable from a very faint star; the appearances of another, Mercury, are so fleeting that few people other than astronomers have ever caught a glimpse of it. This leaves only four planets that are easily and conspicuously seen with the naked eye—Venus, Mars, Jupiter and Saturn.

Venus, as we know, can never be more than 47° along the zodiac from the sun, and therefore never sets more than a few hours after the sun nor rises more than a few hours before it. At its brightest, Venus is rivalled only by the sun and moon—no star and no other planet can approach it in brilliance. Most people have at some time or other been struck by this glittering gem of the twilit sky.

Mars, Jupiter, and Saturn, lying as they do outside the earth, may be seen in any region of the zodiac; indeed, when in opposition they are directly opposite the sun and therefore culminate at midnight. Jupiter is the brightest of the three, and a conspicuous object of the night sky, second only to Venus. Mars is usually less bright, but its conspicuous red tint easily distinguishes it from any other planet. Saturn is the faintest of the three, though still a bright object compared with most of the stars.

In their journeyings round the zodiac two or more planets often pass quite close to one another on the star sphere; such conjunctions are beautiful phenomena, especially when observed with binoculars. Frequently the planets and the moon come into conjunction with each other, and these too should be observed with whatever instrumental aid is available; occasionally the moon passes directly between the terrestrial observer and one of the planets, a phenomenon known as an occultation. These are fascinating to watch with binoculars, particularly when the planet disappears behind or reappears from the dark side of the moon. Regular watchers of the skies will not need to refer to almanacs for warning of conjunctions, since they will—some days or even weeks beforehand—have watched the two bodies drawing steadily nearer to one another. On the other hand it is advisable to be forewarned of occultations by keeping an eye on the almanac.

Finding the Planets

When Venus is at its brightest it is an absolutely unmistakable object. So, too, are Jupiter, Mars and Saturn once the amateur has become familiar with the constellations and knows the positions of the brighter stars. For a bright 'star' lying somewhere on the zodiac where it is known that no star is situated must be one of the planets, and its colour and brightness will usually tell the experienced observer which it is. Alternatively, of course, if their exact positions are discovered from an ephemeris such as the Nautical Almanac and then plotted upon a good star map,[1] a glance from the map to the sky will show whether or not the planet is above the horizon at the time of observation, and in the former event its identification will be a matter of seconds. In the case of Uranus, which to the naked eye is indistinguishable from a faint star, such a procedure is essential. Even when its position has been plotted on the map and the binoculars are directed towards the correct area of the sky, several faint 'stars' may be seen in about the predicted position of the planet and it will be impossible to decide which is Uranus. One thing, however, will give away its identity and clearly distinguish it from the stars—its motion. If the observer notes carefully the position of the suspected planet in

[1] Such as A. P. Norton's *Star Atlas and Reference Handbook* (Gall & Inglis)

relation to the nearby stars, perhaps making a simple map of the region, he will notice after a night or two that one of the faint points of light has slightly altered its position in relation to the others. This must therefore be the planet.

Mercury certainly will not be seen by accident, as Jupiter or Venus may easily be. The amateur who wishes to catch a glimpse of the Messenger of the Gods before he sinks below the horizon or is lost in the brightening glow of dawn must scan the eastern horizon shortly before sunrise during the autumn, or the western sky some twenty minutes after sunset during February and March, or thirty to forty minutes after sunset during April and May. Mercury will then be visible for about one and a half hours before setting, its distance from the sun at maximum elongation varying from 18° to 29°.

Neptune, though never visible to the naked eye, is within the reach of binoculars, and the amateur will be able to spot it by the same method as that already detailed for the locating of Uranus. It will be more difficult, however, in proportion as Neptune is fainter than Uranus and moves more slowly across the background of the stars.

Observing the Planets with Binoculars

Mercury, though found more easily, and observable over a longer period with the aid of binoculars than with the unaided eyesight, cannot be called an interesting object for binocular study. It does not present a sensible disc, and therefore the phases are not discernible, though this does not detract from the thrill of viewing a brother planet which few people—not even the great Copernicus—have ever seen.

Venus, on the other hand, does present a sensible disc in good binoculars when near inferior conjunction (its angular diameter is then about 60″) and the crescent phase can be distinctly seen. When at its brightest, Venus is one of the most severe optical tests to which an object glass or binoculars can be subjected—none but the best will give a clear, symmetrical, colourless image without any trace of fuzziness or 'flare'.

Mars is a beautiful object in binoculars, with its deep red tint, much accentuated when it is near the horizon and especially remarkable when it is in conjunction with the moon, Saturn or Venus and the colours of the two bodies can be directly compared. The disc is not large enough to show any detail with binoculars,

Jupiter, though four times as distant from the sun as Mars, is such a giant planet that it presents to terrestrial observers a larger disc than any of the other planets,[1] clearly visible in good binoculars. Such an instrument should also show the four brighter and larger satellites of Jupiter when they are well placed for observation. But owing to the glare of the planet, the two innermost satellites will be invisible in most binoculars unless a wire is fixed across the focal plane, behind which the planet can be hidden. The temporary invisibility of a satellite may be due to one of several causes: it may be so close to Jupiter that it is hidden in the latter's glare; it may be in occultation (hidden by the body of Jupiter); it may be in transit across the face of the planet; or suffering eclipse in Jupiter's shadow. It is fascinating to note the changing positions of these minute specks of light from night to night or even in the course of a single evening. The Nautical Almanac gives details of their positions, but it must be borne in mind that the sketches of Jupiter and its four moons there given represent their appearance in an inverting telescope; for use with binoculars they must be turned upside down.

The rings of Saturn cannot be seen with binoculars. The giant satellite Titan may just be glimpsed with good glasses when it is in the most favourable position, far from the planet.

With the equipment we are using, Uranus and Neptune are indistinguishable from faint stars, and their true nature is only revealed by their motion. Tracking down and finally identifying these remote and inconspicuous members of the sun's family of planets produces a distinct sense of personal achievement, and it is interesting to watch their stately progress among the stars of the celestial background.

Pluto, as has been said, is invisible in all but the largest telescopes.

[1] Except Venus when nearest the earth.

Chapter 5
THE MOON

Size and Distance

The moon revolves about the earth at a mean distance of 239,000 miles, completing one revolution every 27 days 8 hours. Its diameter is 2,160 miles and it is larger in comparison with the planet about which it revolves than any other satellite in the Solar System; in this respect the moon and the earth are more like twin planets (the moon is not very much smaller than Mercury) than planet and satellite. Our large telescopes reveal objects on the lunar surface no larger than St. Paul's Cathedral.

Rotation

It is widely realized that the moon always turns the same hemisphere towards the earth, for its pattern of dark markings never changes. From this it might be concluded that the moon does not rotate on its axis. This is not so, however. A lunar observer whose line of sight lay past the earth would, in the course of one month's watching, find the entire circle of the heavens pass before his eyes. In other words he has rotated on his axis once, though this fact is liable to be disguised by the fact that during this time the moon has also revolved once about the earth. Thus the moon rotates on its own axis and revolves about the earth in the same period. It follows from this that we can never see more than half of the moon's total surface,[1] the 'far side' being for ever hidden from us.

[1] Actually we can see a little more than half, owing to a rocking motion of the moon's axis, known as libration.

THE MOON

Physical Conditions

The telescopic appearance of the moon—the total absence of twilight, the jet black shadows and glaring high-lights, and in general the crystal-clear definition of its surface features—indicates that it has no atmosphere. This is confirmed by the spectroscope. If the moon has any atmosphere at all, its density cannot exceed one-hundred thousandth of the earth's—which for most purposes is indistinguishable from no atmosphere.

Our own atmosphere contains much water vapour which acts as a blanket for the earth, conserving its heat during the night (a cloudy night is warmer than a clear one) and protecting it from the direct heat of the sun during the day. But the moon has no such protective medium interposed between it and the sun, with the result that its temperature must oscillate much more violently than the terrestrial. The probable figures for the centres of the sunlit and unilluminated hemispheres are 120° C. and −150° C.

Given the former high temperature, and at best an almost non-existent atmospheric pressure, any liquid water that might once have existed on the moon must long ago have evaporated and been dissipated into space.

Summarizing, we may say of the moon; no water, extremes of temperature, negligible atmosphere. Is it any wonder, then, that the moon is always described as a dead world?

Lunar Eclipses

We saw in Chapter 3 how, owing to the inclination of the moon's orbit to the earth's, a solar eclipse does not occur at every new moon. For the same reason, a lunar eclipse—when the moon passes through the earth's shadow—does not occur at every full moon.

Lunar eclipses are always predicted in almanacs and are wonderful spectacles for observation with binoculars. The earth's shadow slowly sweeping across the lunar landscape is a scene of never palling grandeur. The eclipse may be either partial or total, depending on how nearly the moon passes through the centre of the earth's shadow. The totally eclipsed moon is often a beautiful sight, for some of the sun's light is refracted by the earth's atmosphere *into* the shadow and, falling upon the moon, faintly illuminates it with a dusky red light, so that the moon hangs like a copper shield against the deep blue of the night sky.

The Moon's Surface

Since the moon has no 'weather', its hills and mountains have been subjected to none of the processes of erosion that have smoothed and reduced our terrestrial hills from their primordial volcanic outlines. The moon has nothing resembling the gently rolling Downs of Sussex or the undulating uplands of Wiltshire. The greater part of its surface is rugged and barren. However, it is not commonly appreciated that the Moon has virtually no slopes steeper than 20°. Towards the north of the disc lie a number of great plains, darker in tint than the mountainous regions. These plains, or maria (singular: mare), are the dark patches visible to the naked eye.

The most characteristic features of the lunar landscape as revealed by telescope or binoculars, however, are the crateriform objects. These are circular depressions in the surface of the moon, anything from 1 to 150 miles in diameter. All the larger ones have massive ramparts round their periphery, and often a central mountain mass. These ramparts tower many thousands of feet above the floor of the ring.

Lastly there are the bright rays. These are brilliantly white or yellowish streaks which radiate in all directions from some of the brightest craters (notably Tycho, Copernicus, and Kepler). Under a low sun they are invisible, but at Full they contribute largely to the brilliance of the moon. They may be as much as twenty miles in width, and some of the longest run—uninterrupted by mountain ranges, walled plains or whatever surface irregularities they may encounter—for more than 2,000 miles. They cast no shadows and therefore appear to be a coloration of the ground itself rather than topographical features. But why they are only visible under a high angle of illumination is unknown, as also is their exact nature.

The most important lunar surface features, therefore, are:

The Maria: relatively smooth plains, dark in colour.

Mountains: separate mountain ranges (up to about 30,000 feet), isolated peaks, and large mountain masses similar to that of Switzerland.

The Crateriform Objects: from minute pits a few hundred feet across, and hardly visible in the largest telescopes, to gigantic walled plains 150 miles in diameter.

The mysterious systems of Bright Rays.

Examples of all these features can easily been seen with binoculars and are described in the list at the end of the chapter.

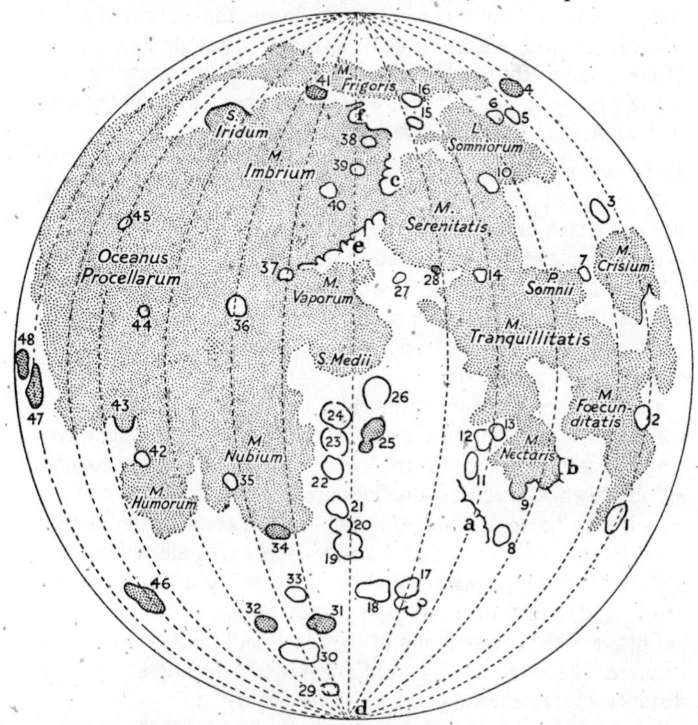

FIGURE 9

Position of terminator shown, from left to right, at the following lunar ages: 13d, 12d, 26d, 10d, 9d, 23d, 7½d, 22d, 6d, 20d, 19d, 3d, 17d.

Earthshine

We can see the moon only because light from the sun is reflected from it into our eyes. But in the same way solar light is reflected from the earth on to the moon. Thus the crescent of the new or old moon is directly illuminated by the sun, but light reflected from the earth also falls upon the dark part of the lunar disc and thence is reflected back into our eyes, rendering this region dimly visible to the terrestrial observer. This phenomenon is known variously as Earthshine, 'The Old Moon in the New Moon's Arms' and the Ashy Light. At times it is a beautiful effect in the evening sky, but with binoculars it is a thousand times more so, a surprising amount of the detail of the 'dark' part of the lunar disc being visible; the maria will easily be made out, while some especially bright craters such as Aristarchus (see p. 73) may also be spotted.

Radar and the Moon

It has been known for more than half a century that radio waves—which differ from light only in having longer wavelengths—are capable of being reflected. It is the reflection of radio waves by electrified layers in the upper atmosphere that makes possible the reception in England of a programme transmitted from Australia: for radio waves can only travel round corners if they have something to bounce off.

In its essence, a radar equipment consists of a radio transmitter which sends out a beam of evenly spaced pulses on a short wavelength (commonly 4 to 6 metres). When the radio waves encounter an object whose electrical qualities are different from those of the air—it may be an aeroplane, a cliff, a cloud even, or the ionized gases lying along the path of a meteor—they are reflected from it.

These incoming waves are then collected by an ordinary receiver, and used to operate a device which measures the time interval between the transmission of the signal and their arrival at the receiver. The speed at which radio waves travel being known, the distance of the reflecting object can be calculated.

In practice, no calculation is involved, the time-measuring device (a cathode ray tube) being calibrated in the equivalent units of distance—miles, thousands of yards, or other convenient unit.

THE MOON

Perhaps the most spectacular feat of radar in astronomy was achieved at the Evans Signals Laboratories, in the United States, just after the war. In January 1946 an ordinary army radar set was used to transmit a 7°-wide beam at the moon. Echoes were received 2½ seconds after the pulses were transmitted . . . for the first time in history, contact had been established with another world.

Accurate ranging was not possible with the equipment then available, but it was nevertheless a tantalizing foretaste of things to come. For radar has one telling advantage over the trigonometrical parallax method[1] which, up to the present, has been the astronomer's only method of measuring distances within the Solar System: whereas the latter becomes less and less accurate as the distance which is being measured increases, the inaccuracy margin of radar measurements is constant, whatever the distance. The linear error in the derived range of a planet millions of miles away will be no greater than that in the range of an aeroplane flying overhead.

A parallel development that may be expected is the detailed mapping of the contours of the moon's surface by radar. Combined with telescopic observation, this will give us more precise knowledge of the lunar topography than we have of large areas of the earth's surface.

Observing the Moon

To an airman, looking down on the earth far beneath, no difference can be distinguished between hilly country and flat; both look like a smooth plain. So, too, when we on earth look at the full moon, we gaze down upon mountains and valleys, ravines and great crater rings which are illuminated from directly overhead; they consequently cast no shadows and appear as flat and uneventful as the airman's view of the Lake District. To the uninitiated it comes as a surprise that when the moon is full, to the naked eye the glorious Queen of the Heavens, then telescopically she is a dowdy and uninteresting old woman.

[1] See p. 84.

The line of demarcation between the illuminated and the dark sides of the lunar disc is known as the terminator, and the position of this terminator (which of course crosses the disc twice in each lunation) is of paramount importance to the observer with binoculars or telescope. When on or near the terminator a mountain, say, casts impenetrably black shadows against which it is vividly silhouetted, every detail of its structure standing out starkly: a week or two later it is illuminated by a high instead of a slanting sun and it may appear as a bright featureless spot or even be totally invisible. When studying the formations described below and figured on the map, observe them when near the terminator. Because of the decisive importance of this factor in lunar observation, the position of the terminator at intervals of a few days is shown on the map, so as to give the beginner some idea of when during the month any particular district of the lunar landscape is best observed.

Short List of Objects for Binocular Study

Mare Crisium (Sea of Crises). Most conspicuous of the maria on account of its distinct outline and isolated position; clearly seen by the naked eye. It measures roughly 350 by 300 miles (area, 78,000 square miles). Jutting out 60 miles into the Mare from the south is the *Promontorium Agarum*, a mountain mass rising to some 11,000 feet above the surface of the plain. Best observed when the moon is 15–16 days old, or a day or two after new.

Mare Foecunditatis (Sea of Fertility). A large, ill-defined, dark plain lying to the south of the Mare Crisium. On the north it opens into Mare Tranquillitatis. Area about 160,000 square miles.

Mare Nectaris (Sea of Nectar). One of the smaller maria, opening into the southern reaches of Mare Tranquillitatis. The Pyrenees Mountains form its western border.

Mare Tranquillitatis (Sea of Tranquillity). One of the largest of the Maria. A 'bay' in its north-west shore is called *Palus Somnii* (Marsh of Sleep). It is continuous with Mare Serenitatis on the north and Maria Nectaris and Foecunditatis on the south.

Mare Serenitatis (Sea of Serenity). One of the most prominent of the great plains. A journey round its 'shore' would take the traveller some 1,850 miles (area about 125,000 square miles). On its north side the minor plain known as *Lacus Somniorum* (Lake of Dreams) opens into it. At Full, binoculars will show a narrow bright ray running across the mare from the direction of Menelaus; actually it emanates from Tycho, two thousand or so miles distant.

Mare Humorum (Sea of Humours). A small plain, 260 miles from north to south, 280 from east to west, area about 50,000 square miles. To its west lies *Mare Nubium* (Sea of Clouds), which like it opens into the great *Oceanus Procellarum* (Ocean of Storms).

Sinus Medii (The Central Bay). An indistinct greyish area lying at the mean centre of the moon's visible hemisphere. Roughly 13,000 square miles.

Mare Imbrium (Sea of Showers). Largest of the maria; 750 by 670 miles. Bounded on the west by the Alps and the Apennines; on the north by the mountainous country separating it from *Mare Frigoris* (Sea of Cold); to the east it is continuous with *Oceanus Procellarum*. In its northern shore lies the conspicuous bay, *Sinus Iridum* (Bay of Rainbows). It looks as though carved out of the mountainous country that bounds the mare in this region, and at points the 135 mile line of cliffs towers 15,000 to 20,000 feet above the plain. The cape at each end of the bay (Laplace and Heracleides) can be made out with binoculars under suitable illumination.

Petavius (1)[1] A great walled plain, 100 miles in diameter. Its ramparts rise to a height of 10,000 feet on the east. Disappears completely under high illumination.

Langrenus (2). Diameter, 80 miles. The interior and the large central mountain are very bright. Ramparts up to 10,000 feet.

Cleomedes (3). Diameter about 80 miles. Walls up to 9,000 feet on the west.

Endymion (4). Diameter, 78 miles. Walls 15,000 feet on the west. Floor unusually dark.

Atlas (5). The twin and close neighbour of Hercules (q.v.). Dia-

[1]Numbers or letters in brackets refer to the Lunar Map.

meter 55 miles. Walls 11,000 feet on the north.

Hercules (6). Diameter about 45 miles. Ramparts to 10,000 feet.

Proclus (7). One of the brightest formations on the moon, its ramparts at Full showing in binoculars as a brilliant white spot although its diameter is less than 20 miles.

Piccolomini (8). Situated at the southern end of the Altai Mountains. Diameter, 55 miles. Height of walls, 9,000 feet to 14,000 feet.

Altai Mountains (a). A mountain range running north-east from Piccolomini for a distance of 300 miles. Its western face is a precipitous cliff; individual peaks reach a height of 13,000 feet.

Fracastorius (9). Partially ruined, and appears as a bay in the southern shore of Mare Nectaris. Diameter about 60 miles.

Pyrenees (b). A mountain range, well seen when near the terminator, which forms the western border of Mare Nectaris. To the north it reaches a height of 12,000 feet.

Posidonius (10). A walled plain, 60 miles in diameter, situated on the promontory jutting out between Lacus Somniorum and Mare Serenitatis. Ramparts reach a maximum height of 6,000 feet.

Caucasus Mountains (c). A mountain mass which forms the north-east edge of Mare Serenitatis. Peaks from 12,000 feet to 18,000 feet. Note its shadows lying across the mare about first Quarter.

Catharina (11). The southernmost member of a row of three walled plains lying closely north of the Altais. Diameter about 70 miles. Ramparts, very irregular, rise of 16,000 feet.

Cyrillus (12). Contiguous with Catharina. Diameter, 65 miles.

Theophilus (13). The third member of this group. Diameter about 65 miles. At their highest point the ramparts are 18,000 feet above the floor of the ring.

Plinius (14). A ring plain 32 miles in diameter, lying on the northern edge of Mare Tranquillitatis.

Eudoxus (15). Lies north of the Caucasus Mountains, between them and Mare Frigoris. Diameter, 45 miles. Ramparts, 11,000 feet.

Aristoteles (16). The twin of Eudoxus, some 50 miles to the north of which it lies, on the shore of Mare Frigoris. Diameter, 50 miles; height of ramparts, 11,000 feet.

Maurolycus (17). One of the many fine walled plains in the region of the south pole which have to be observed when near the terminator. Diameter, 150 miles. Maximum height of walls, 18,000 feet. At Full it is invisible.

Stöfler (18). Resembles, and lies closely east of, Maurolycus. Irregular in shape and partially ruined. Diameter about 150 miles. Ramparts, 12,000 feet. Invisible at Full.

Walter (19). The southernmost member of a line of six great walled plains on the moon's central meridian (Nos. 19–24). Irregular in shape, from 80 to 100 miles across. Ramparts up to 10,000 feet.

Regiomontanus (20). Also irregular in shape; greatest diameter about 80 miles.

Purbach (21). Diameter, 70 miles. Ramparts, 8,000 feet.

Arzachel (22). Diameter, 65 miles. Massive ramparts reaching heights of from 10,000 feet to 13,000 feet.

Alphonsus (23). 80 miles in diameter. Walls low (5,000 feet to 9,000 feet) but very massive, being in places 15 miles wide.

Ptolemaus (24). The finest and most northern of the six. From 90 to 115 miles across. Walls, 13,000 feet.

Albategnius (25). A fine walled plain with a remarkably dark floor. Diameter, 80 miles. The massive ramparts, in places nearly 20 miles thick, rise to a height of 14,000 feet.

Hipparchus (26). Forms an interesting pair with Albategnius. A giant walled plain, somewhat irregular in shape, measuring 90 miles from east to west.

Manilius (27). A smaller formation than the foregoing, situated to the west of Mare Vaporum. Though only 25 miles in diameter ts unusual brightness renders it clearly visible with binoculars under all illuminations. Ramparts, 7,000 feet high.

Menelaus (28). An even smaller crater (diameter, 10 miles) lying on the shore of Mare Serenitatis at the foot of the Haemus Mountains (8,700 feet). Its ramparts rise about 6,000 feet above the floor. These are very bright, and the crater is conspicuous at Full despite its small size.

Leibnitz Mountains (d). A great range situated on the moon's limb at the south pole, only partially visible to terrestrial observers. It contains some of the highest peaks on the moon (30,000 feet). Well seen both at Full and during the first Quarter along the southern cusp.

Moretus (29). Lies close to the south pole and must be observed on or near the terminator, since it disappears at Full. Diameter about 75 miles. Ramparts massive, 9,000 feet. Notable in that its central mountain peak is one of the highest of this type of mountain on the moon (about 7,000 feet).

Clavius (30). Another gigantic walled plain in the vicinity of the south pole—the largest on the moon: diameter, 140–150 miles. Walls 12,000 feet high. Invisible at Full.

Maginus (31). About 110 miles in diameter. Ramparts, 14,000 feet. The floor is much darker than the surrounding country. Study this vast enclosure when it is near the terminator and reflect on the seemingly impossible fact that for about three days at Full it is entirely invisible.

Longomontanus (32). Resembles Maginus. Dark floor, diameter about 90 miles, ramparts up to 13,000 feet, invisible at Full.

Tycho (33). The centre of the largest and most brilliant of the ray systems. It is owing to the glaring white brilliance of all this region near Tycho under high illumination that so many of the greatest formations similar to those just described are invisible at Full. So bright is Tycho at this Quarter that it can be discerned by the knowing eye without optical aid, although only 55 miles in diameter.

Pitatus (34). A dark-floored ring lying on the southern edge of Mare Nubium. Diameter about 50 miles.

Bullialdus (35). A bright ring plain situated in Mare Nubium. Diameter, 38 miles. It lies on one of the bright streaks from Tycho.

Copernicus (36). One of the finest of the lunar rings. Diameter, 56 miles; ramparts, 12,000 feet high. Lies on a bright patch in Mare Imbrium. Centre of a system of bright rays, and both it and its surroundings are very bright at Full. Best studied, therefore, near the terminator.

Eratosthenes (37). Lies between Copernicus and the Apennines. Diameter, 38 miles. Ramparts from 10,000 feet to 15,000 feet. Conspicuous central mountain. Invisible with binoculars at Full.

Apennines (e). Perhaps the grandest of the lunar mountain ranges. It towers 18,500 feet above the smooth expanse of Mare Imbrium, whose south-west border it forms for a distance of 500 miles. Even in binoculars it is a fine and arresting spectacle when near the terminator, its jet black shadows falling 100 miles across the plain. Most of its intricate detail is unfortunately invisible with such an instrument: the telescope shows it to be scored by great valleys and ravines, while lofty peaks rear earthwards to heights of from 15,000 to 20,000 feet. More than 300 individual peaks have been mapped.

Cassini (38). The first of three ring plains situated at the western end of Mare Imbrium (Nos. 38–40). Diameter, 36 miles. Low ramparts.

Aristillus (39). Diameter, 35 miles. Ramparts 8,000 feet on the east, 11,000 feet on the west.

Archimedes (40). Diameter, 50 miles. Low (about 4,000 feet) massive walls.

Plato (41). A fine, distinct ring plain lying back from the north shore of Mare Imbrium. Notable for the dark tint of its floor. Diameter, 60 miles. Walls up to 5,000 feet.

Alps (f). A great mountain mass forming the north-west border of Mare Imbrium between the crater rings Plato and Cassini. Its highest peak, Mt. Blanc, towers 12,000 feet above the plain. The Alps are cleft by a great valley, 80 miles long and from 3½ to 6 miles wide, which connects Mare Imbrium with Mare Frigoris.

Gassendi (42). A ring plain, 55 miles in diameter, lying on the north-east shore of Mare Humorum. Walls from a few 100 to 10,000 feet high.

Letronne (43). A partially ruined ring, appearing as a bay in the south shore of Oceanus Procellarum (cf. Fracastorius). About 50 miles across.

Kepler (44). A small (diameter, 20 miles) bright crater, situated in Oceanus Procellarum. It lies on a bright patch which is very conspicuous at Full and is the centre of one of the great ray systems.

Aristarchus (45). Like Kepler, an isolated bright crater (probably the brightest spot on the moon) situated in Oceanus Procellarum. Quite featureless though very conspicuous at Full, it is best observed close to the terminator. Bright rays radiate from it under high illumination. Diameter, 30 miles. Walls, 7,000 feet.

Schickard (46). We now come to the eastern limb of the moon. Schickard is a fine walled plain, 135 miles from east to west. Ramparts from 4,000 feet to 10,000 feet.

Grimaldi (47). A vast walled enclosure, measuring about 150 by 130 miles. Ramparts, partially ruined, average some 4,000 feet in height. Its floor is perhaps the darkest spot on the moon: sharp eyesight can detect it without optical aid.

Riccioli (48). Lies between Grimaldi and the limb. Also dark, but less so than Grimaldi. Diameter, 106 miles.

Chapter 6

ASTEROIDS, COMETS AND METEORS

The Sweepings of the Solar System

The sun, the nine planets and their thirty-one satellites do not represent the whole of the Solar System, though these bodies are admittedly the most important if importance is to be measured in tons. Scattered throughout the system are millions upon millions of meteoric particles, material which we may perhaps regard as the detritus of creation—material which never condensed to form major bodies like the planets. These particles vary in size from bodies comparable with the smaller satellites down to minute specks of dust weighing but a small fraction of a gram. They are of three distinct types: asteroids, comets, and meteors.

The Asteroids

If you look at column 3 of the Table on p. 119 you will notice that there is an abnormally wide gap between the orbits of Mars and Jupiter. The existence of this gap is clearly brought out by what is known as Bode's Series. If each term in the series

0 1 2 4 8 . . .

is multiplied by 3 and then added to 4, we get

4 7 10 16 28 . . .

As the Table below shows, these numbers correspond very closely with the distances of the various planets from the sun, except (*a*) towards the outer edge of the Solar System, and (*b*) that there is no planet corresponding with 28:

Bode's Series	*Planet*	*Distance from Sun (Earth = 10)*
4	Mercury	3·9
7	Venus	7·2
10	Earth	10·0
16	Mars	15·2
28		
52	Jupiter	52·0
100	Saturn	95·4
196	Uranus	192
388	Neptune	301
772	Pluto	395

Towards the close of the eighteenth century it was suggested that a still undiscovered planet might occupy this gap. A search was initiated and not one but *several* small planetary bodies were quickly discovered; since that time the discovery of these 'planetoids' or asteroids has proceeded apace, being greatly accelerated by the application of photography to astronomy; today between one and two thousand are known. Though these lie for the most part between the orbits of Mars and Jupiter, odd ones are scattered over the whole tract from the orbit of Mercury to that of Saturn. It has been estimated that there must be something like 40,000 asteroids bright enough to be seen with our largest telescopes, but that their combined mass probably does not exceed 2 per cent of the moon's.

The diameters of the four largest are 490 miles (Ceres), 304 miles (Pallas), 248 miles (Vesta), and 118 miles (Juno). The smallest yet investigated have diameters of the order of one mile and it is probable that there is no sharp line of demarcation between the asteroids and meteors that could be held in the hand. Vesta, though not the largest, is the brightest of the asteroids and the only one that is ever visible to the naked eye. It is certain that not even the largest possess atmospheres, and it is likely that they are nothing more than barren lumps of rock. Many of them vary in brightness, probably indicating that they are not spheroids like the planets but are irregular in shape.

Many of the asteroidal orbits are very odd according to planetary standards. They are not confined to the zodiac—the greatest inclination of an asteroidal orbit to the ecliptic yet discovered is 48°. Furthermore, the orbits are often so elongated or eccentric that the solar distances of the asteroids vary enormously as they revolve about the sun. Thus an asteroid discovered in 1932 lies eight million miles within the orbit of Venus when at perihelion (nearest the sun) and at aphelion (when furthest from the sun) sixty-one million miles outside Mars! Its orbit thus lies athwart those of three major planets. This asteroid can approach within three million miles of the earth—our nearest neighbour after the moon. A tiny asteroid discovered as recently as June 1949 approaches the earth to within four million miles, and when nearest the sun is actually within the orbit of Mercury; it is probably less than a mile in diameter, and takes only six weeks longer than the earth to complete one revolution round the sun. Another 1932 discovery, known as the Reinmuth planet, comes within ten million miles of the earth, and Eros (a larger body than either of these, having a diameter of about twenty miles) within some fourteen million miles.

Comets

The advent of large and brilliant comets has from time immemorial been regarded as presaging public woes, or death and calamity in high places. And conspicuous comets are indeed remarkable apparitions: in some cases they are bright enough to be visible in broad daylight, their immense tails streaming half-way across the dome of the sky. Such comets, of rare occurrence, are seen to consist of a minute star-like nucleus, a misty hood or coma that surrounds it, and a great curved tail directed always away from the sun. The greater number of comets, however, are small and faint objects only visible with instrumental aid; these are not normally differentiated into a clearly defined head and tail but appear simply as structureless hazy spots.

The last great comet to arrive at perihelion was Halley's, which returns every seventy-five years. On its return in 1910 it was visible to the naked eye for four and a half months, its length then being some twenty-eight million miles.

Their Behaviour and Orbits

Comets revolve around the sun, in common with all the other members of the Solar System, but their orbits differ from those of planets in several important respects. In the first place they may be inclined to the ecliptic at any angle up to 90°; comets, therefore, are not confined to the zodiac but may appear in any part of the heavens. Secondly, they are highly eccentric—when at perihelion a comet may be only a few hundred thousand miles from the sun's surface, and at aphelion millions of miles beyond Pluto.

The shortest known period of revolution is 3·3 years, while the longest are several centuries or even millennia. Of these great periods of time the comet is only visible to terrestrial observers for a comparatively few months while the comet swings round the sun before receding once more to the furthermost confines of the Solar System.

The tail of a comet always points away from the sun, even when the comet is moving from perihelion to aphelion and therefore proceeding tail foremost. It thus appears that some repellent agency is situated in the sun and that this always drives the tail away from the head. This repulsive force is the pressure (infinitesimally small except in the sun's immediate vicinity) that light exerts upon any surface or particle which it strikes.

Comets are discovered at an average rate of five per annum, though as many as fourteen were discovered in 1947, a particularly good year. The vast majority of these are faint, tailless objects, only to be seen with telescopes or binoculars. About two-thirds of them are new comets never observed before, the remainder being previously known comets returning to perihelion.

What is a Comet?

Comets are the largest bodies in the Solar System: the coma of a well developed comet may measure several hundred thousand miles across, and the tail many millions of miles in length. Yet the mass of even the greatest comets is too minute to be detected and cannot exceed about 1/100,000 of the earth's. When the tail or even the head of a comet passes between us and a faint star, the latter suffers absolutely no diminution of its brightness. In 1910 the earth passed through the tail of Halley's comet without any effect whatsoever being observed, and on the same occasion the comet's head passed in transit across the sun's disc but was invisible in the most powerful telescopes.

It is thought that the head of a comet consists of a widely spaced swarm of meteoric bodies, and the coma and tail of highly rarefied gaseous or dusty material which is expelled from the head near perihelion by the light pressure of the sun. Spectroscopy shows that comets, at any rate when in the vicinity of the sun, shine partly by reflected sunlight and partly by virtue of their own radiation.

Comets and Meteors

In 1826 Biela discovered a comet with a period of 6·6 years. At the 1846 return the comet was seen to have split into two. At the next apparition in 1852 these two components were 1½ million miles apart. The comet was never seen again, but on subsequent occasions when it should have returned there was instead a brilliant display of meteors, and it has been discovered that the orbit of this meteor swarm is identical with that of Biela's comet. This argues a very close connexion between the two classes of body, and substantiates our conception of the nature of the heads of comets. A similar identification of cometary and meteoric orbits has been made in a number of other instances.

Observing Comets

Nowadays—with the universal professional use of photography in astronomy, and with amateurs scattered over the globe who systematically search for new comets with the appropriate equipment—there is little likelihood that the casual watcher of the skies will be allowed to claim a cometary discovery. Every year, however, comets are discovered which are within the reach of small telescopes and binoculars, and the location of these and their subsequent observation as they move rapidly across the background of the stars, brightening up to perihelion and then slowly fading into invisibility, is a fascinating occupation. Details of the position, rate and direction of movement, and brightness of newly discovered comets are given in such periodicals as *Nature, The Observatory* and the *Journal of the British Astronomical Association.*

Meteors

At some time or another everyone has seen a meteor, looking like a star that has slipped its moorings and is tumbling down from the firmament. These transient points of light are in reality small fragments of rock or metal ore, weighing usually only a few milligrams, which in the course of their journeyings round the sun are drawn into the earth's gravitational field. Hurtling through the atmosphere at velocities which are commonly of the order of 20 to 30 m.p.s. they are heated to incandescence by the resistance of the air, and all but the most massive are completely vaporized long before they can reach the earth's surface. Meteors are first visible at a height of about a hundred miles and the majority are burnt out while still more than thirty miles above us.

Meteor Swarms

When the flight of each meteor seen during a night's observing is plotted on a star map it will often be found that the tracks appear to diverge or radiate from a single point on the star sphere. This point is known as the radiant, and the meteors which, when caught in the earth's atmosphere, radiate from such a point, are said to constitute a meteor swarm. When a shower of this sort occurs it means that the earth is passing over the point at which its own orbit intersects that of the meteors, which like the earth are revolving about the sun. These meteors will of course be travelling in space in more or less parallel paths, and their apparent divergence from the radiant is merely an effect of perspective.

Every year when the earth reaches the point of intersection of the two orbits a shower will be observed, since there are stray meteors scattered round the entire orbit. But the simultaneous arrival of the earth and the swarm itself at the intersection of the two orbits may only occur at intervals of several years. Thus there was an exceptionally fine display of Leonid meteors every thirty-three years. Each swarm—of which hundreds are known, although even the few conspicuous ones do not usually yield more than a few score meteors per hour to any one observer—is named after the constellation in which the radiant lies; if there is more than one radiant in any constellation the designation of the bright star nearest to the radiant is prefixed, e.g. the δ Aquarids, κ Cygnids, etc., indicating that the radiants lie near δ Aquarii and κ Cygni respectively.

Meteors and Radar

Since 1946 a powerful new instrument for the observation of meteors has been in the hands of astronomers—radar. Owing to the high temperature generated by a meteor, the atoms of the atmosphere in its path are ionized; that is to say, one or more of their orbital electrons are stripped from them. And it has been found that this stream of ionized atoms and free electrons is capable of reflecting a radar beam.

Every time a meteor crosses the radar beam a bright 'splash' appears on the screen of the cathode ray tube of the receiving set, and from its position the distance of the meteor can be read straight off the range scale—just as though it were an enemy aircraft. It is possible at the same time to determine the meteor's direction—the bearing and altitude of its beginning and end points—so that its position in space can be completely plotted.

Even as an auxiliary to the visual observation of meteors radar would have been welcome. But it is more than just that, for it possesses two outstanding advantages over the visual method: it can be used when clouds make visual work impossible, and it can be used just as well by day as by night. The most intriguing and unexpected accomplishment of radar in this branch of astronomy was the discovery in 1947 of a major meteor swarm which not only never had been seen before, but never will be—be-because its meteors occur during the hours of daylight. This system of meteors is one of the richest and most important known: it produces an average of eighty meteors per hour, day after day for fully three months in the year—May to August. Without radar its very existence would never have been suspected.

How many Meteors are there?

It has been estimated that although a single observer would count himself lucky to see as many as eighty meteors per hour even when the earth is passing through a swarm, the entire atmosphere of the earth every day sweeps up something like twenty or thirty million meteors bright enough to be seen by the naked eye. If telescopic meteors are included, this number becomes many times greater. Taking figures that must of course be widely speculative, but which probably err on the low side, let us say that 100 million meteors whose average weight is five milligrams enter the atmosphere every twenty-four hours. Then the weight of the earth is increasing through meteoric additions at the rate of about 4,500 tons per annum!

Observing Meteors

A meteor is not in itself an interesting thing to observe—merely a moving dot or streak of light lasting a few seconds. But when swarms are studied by the careful plotting of their individual members on a star map, or when observations are made simultaneously from different stations, many interesting problems are uncovered. The systematic observation of meteors lies entirely in the hands of amateurs, and in this country the work is organized and the results collated by the British Astronomical Association.

Meteor observation requires little besides patience, a good working knowledge of the constellations, and a straight piece of wood about a yard long. The procedure is as follows. Each time a meteor is seen, the observer starts counting to himself: '*Nought* one two three *One* one two three *Two* one two three . . .' and stops when the meteor vanishes; with very little practice the speed of counting can be regulated so that the italicized words mark the meteor's duration in seconds. At the same time the observer holds up his straight-edge so that it lies along the track of the meteor, and notes its position in relation to the stars. With this to aid him, he then rules a line on a star chart to represent the meteor's path, and makes a note of its duration and the time of its occurrence. Any other points of interest—the meteor's colour, if marked; its brightness; whether or not it left a train, etc.—can be noted too. A more accurate method of observing and recording meteors is described fully in my book *Observational Astronomy for Amateurs*.

If any reader feels he would like to carry out work which is not only well within the scope of his qualifications and equipment, but which in addition will be of permanent value to science, he should get in touch with the B.A.A. For those who wish to observe meteors casually, the list of constellations in Appendix 3 contains details of all the more important meteor showers.

Chapter 7

THE STARS

We have already learnt that the stars are great spheres of incandescent gas—objects of the same general type as the sun. When we consider how much brighter the sun is than even the most brilliant star, we will naturally and correctly conclude that the stars are many times more distant from the earth than the sun is. It is possible by various methods to determine the distance of the stars with considerable accuracy.

Finding the Distance of a Star

When an observer's viewpoint moves, near objects appear to change their positions against the background of more remote objects: thus the view from a railway carriage window consists of telegraph poles flashing past the more distant landscape. The nearer an object is to the observer, the greater will be this apparent displacement for a given shift on his part. In fact, the amount of the displacement is inversely proportional to the distance of the object, and the latter may easily be calculated from the former.

Now in a period of six months a terrestrial observer alters his position in space by some 186,000,000 miles[1] (the diameter of the earth's orbit) and one might suppose that this displacement of the observer would result in the nearer stars appearing to alter their positions against the background of the more distant stars. This is actually the case, but owing to the immense distances of even the nearest stars the shifts are extremely minute, and escaped detection until Bessel measured the first stellar distance in 1838. The parallactic shift of 61 Cygni, the star in question, only amounts to

[1] Leaving out of account the motion of the sun itself.

one five-thousandth of the apparent diameter of the moon! This corresponds to a linear distance of 64,800,000,000,000 miles.

The Light Year

Obviously it is most inconvenient to express such enormous distances in units as small as the mile; instead, the 'light year' is employed. Light travels with a velocity of 186,000 miles per second and this distance is accordingly known as a light second. Similarly, the distance that light travels in a year (6,000,000,000,000 miles) is one light year. Thus the distance of 61 Cygni is 10·8 L.Y.—a much more manageable expression.

The *nearest* star, Proxima Centauri, is 4·2 L.Y. distant from the sun; the most distant are hundreds of thousands of light years away.

A Star's Distance from its Luminosity

Unfortunately the parallactic shift becomes too small at distances greater than about 200 light years for this method to be used with any success, and the astronomer is therefore forced to fall back on alternative methods for stars more distant than this. If we can discover the real brightness, or luminosity, of a star we can immediately deduce its distance by a simple calculation. For the apparent brightness of any light source is inversely proportional to the square of its distance from the observer. Since we know how bright the star appears to be, we only require to know how bright it really is in order to deduce its distance.

Stellar Luminosities

These real brightnesses vary enormously as between different stars. In other words, if all the stars were brought to the same distance from the sun they would still appear to be of greatly differing brightnesses.

The most intrinsically brilliant stars have luminosities of the order of 50,000 times that of the sun, though one star, S Doradus, is suspected of being 300,000 times more luminous than the sun. At the other end of the scale we have stars only 1/50,000 as bright as the sun.[1] The difference between these two extremes of lumin-

[1] A star discovered in 1944 is only two-millionths as luminous as the sun.

osity may be likened to, though still even greater than, the difference between a candle and the most powerful searchlight.

Stellar Magnitudes

A glance at the night sky shows that the stars differ among themselves in apparent brightness—differences dependent upon a combination of unequal luminosities and unequal distances. These differences in apparent brightness are measured on the scale of stellar magnitudes, and it is important to realize at the outset that the 'magnitude' of a star is solely an indication of its brightness and has nothing whatever to do with its linear size.

Roughly speaking, the twenty brightest stars in the sky are of the first magnitude and the remainder of the stars visible to the naked eye are divided into five lower magnitudes, the faintest stars discernible on a clear, moonless night belonging to the sixth magnitude. The stars of each magnitude are 2·5 times as bright as those of the next lower magnitude, so that a first magnitude star is about 100 times brighter than one of the sixth magnitude. On the same scale, the magnitude of the sun is –27 and that of the faintest stars shown by the Mt. Palomar 200-inch reflector, about 23.

The Temperatures of the Stars

A star resembles a lump of iron in that its temperature determines it colour: a star or an iron bar at red heat is cooler than one at white or blue heat. Thus we can obtain a rough idea of the temperature of a star just by looking at it, but the detailed examination of its spectrum enables us to reach a much more accurate result.

Stellar temperatures vary over a smaller range than do their luminosities, the limits being (excluding exceptional cases) about 30,000° and 2,000°. The nature of the correlation between colour and approximate temperature is shown below:

Greenish or bluish white	30,000°–19,000°
Yellowish white	11,000°– 7,500°
Yellow	7,500°– 5,000°
Deep yellow or orange	5,000°– 3,000°
Red	3,000°– 2,600°

Below about 2,000° a star's radiation will be largely invisible—it will be hot but not shining. ϵ Aurigae, which is the coolest as well as the largest star known, is of this sort: its temperature is only 1,700° and the greater part of its radiation is in the infra-red, i.e. invisible heat radiations.

Size, Mass, and Density

The range of stellar size is, like luminosity, enormous. At the lower end of the scale lie the dwarfs, bodies comparable in size with the planets; a typical white dwarf, Sirius B, is described on p. 168. At the other end are the giants, the greatest of which have diameters larger than that of the orbit of Mars; indeed, were ϵ Aurigae to be substituted for the sun, then Mercury, Venus, the earth, Mars, Jupiter and Saturn would all lie inside it! Such stars are as much larger than the sun as the latter is larger than the average planet.

Stellar mass, on the other hand, varies within comparatively small limits, the vast majority of the stars being less than ten times and more than one-fifth as massive as the sun. It follows, therefore, that the density of the giants is exceedingly low, while that of the dwarfs is quite fantastically high, for the definition of density is $\frac{\text{Mass}}{\text{Volume}}$. Once these two quantities are known, a quick calculation reveals the fact that the density of a giant star such as Betelgeuse is about equivalent to what a physicist would describe as a tolerably good laboratory vacuum. The mass of a dwarf, on the other hand, is compressed into such a small space that the density is many times greater than that of lead: so tightly compressed, in fact, that one cubic inch of the matter of the dwarf Sirius B would weigh a ton.

The Sun as a Star

It is interesting to see where the sun stands in the stellar hierarchy, and the Table below shows that it is in no respect outstanding and is indeed rather below the average in most instances—a *petit bourgeois*, in fact. The figures are approximate:

	Maximum (normal)	*Sun*	*Minimum (normal)*
Luminosity	50,000	1·00	0·00002
Temperature	30,000°	6,000°	2,600°
Diameter (miles)	400,000,000	864,000	8,000
Mass	10	1·00	0·2
Density	60,000 × water	1·4 × water	equivalent to good 'vacuum'

Variable Stars

There is a special class of star which deserves mention since its members are of great interest to observers with inexpensive equipment such as binoculars. These are stars whose brightness is not steady, but fluctuating, either regularly in recurrent cycles or else without apparent rhyme or reason. Five types of variable may be distinguished.

(i) *Irregular Variables*. These are typically red stars, and the range of variation seldom exceeds half a magnitude or so. Betelgeuse is an example from among the brighter stars. The cause of this type of variation is unknown. Examples of irregular variables, as of all the other types, are given in Appendix 3.

(ii) *Long Period Variables*. Unlike the foregoing, their brightness fluctuates fairly regularly and predictably between definite maxima and minima. The range of the variation is usually several magnitudes, and they are therefore more noticeable than the irregular variables. The periods of these stars (i.e. the interval between successive maxima or minima) lie between several months and several years. The cause of the variation is thought to be the rhythmic pulsation of the star itself.

(iii) *Cepheids*, named after the variable, δ Cephei. The variation proceeds with clockwork regularity, the periods usually being measured in days. Cepheids also are probably pulsating stars. A most peculiar relationship has been discovered to exist between the periods and the luminosities (or intrinsic brightness) of these variables, such that the longer the period the more luminous is the star. This period-luminosity relationship is of the greatest value as furnishing us with an alternative method of determining stellar distances and the distances of star clusters (see pp. 94-95), many of which contain Cepheids. For it is only necessary to measure the period of a Cepheid in order to deduce its intrinsic brightness, when a comparison of this with its observed brightness at once yields its distance. The reason for this relationship is unknown.

(iv) *Eclipsing Variables.* A variable of this type is not a single, isolated star but a binary system consisting of a pair of stars too close to one another to be observed individually. These components revolve about one another in orbits which are seen edge on from the earth, with the result that at regular intervals the two stars eclipse one another. The relative size and brightness of the two stars, their distance apart and the inclination of their orbits to our line of sight are factors which affect the exact nature of the light variation, but all eclipsing variables have light curves of the same general type.

(v) *Novae*, or Temporary Stars. These are faint stars that suddenly brighten to a brilliant maximum and then fade again, very much more slowly and with many minor fluctuations of brightness, to near their original magnitude. The light output during the short period of increasing brilliance (usually a matter of days) is stupendous. Nova Aquilae 1918, for example, increased its brightness 40,000 times in four days, while the 1885 nova in the Andromeda Nebula (see p. 138) emitted more light in six days than the sun emits in a million years! Up to the present, some fifty novae have actually been observed at the time of their occurrence (others have been discovered 'posthumously' on photographic plates). Of these, perhaps the most famous, and certainly the brightest, was that observed by Tycho Brahe in 1572; this was visible to the naked eye for the unusually long period of eighteen months, and at maximum was brighter than Venus and clearly visible by day. Now it is fainter than the twelfth magnitude. The cause of these cataclysmic explosions is unknown, but explosions within the interior of the star they almost certainly are.

Binary Stars

In twc different instances we can discern that an apparently

single star is in reality a pair of stars which are not individually visible. The first we have already noticed—eclipsing variables. The second is the type of star known as a spectroscopic binary, which is nothing more than an eclipsing variable which does *not* vary because the plane of the components' orbits does not pass near the earth, and consequently no eclipses are observed. The spectrum of such a star exhibits a periodic shifting or doubling of its lines, indicating that the star is alternately approaching and receding from the earth. This in turn indicates that it is travelling in an orbit about an invisible companion.

If, however, the components are much farther apart than those of either spectroscopic or eclipsing binaries, it becomes possible to detect them by direct vision: what appears to the naked eye to be a single star is seen in the telescope or binoculars to be a close pair of stellar points. Whereas the revolution periods of spectroscopic binaries are short (in some cases as short as one day), the components being relatively close to each other, the periods of visual binaries are not shorter than several years and as often as not are several centuries or even millennia. Many binaries which repay scrutiny with binoculars are listed in Appendix 3.

The Life and Death of a Star

Until about 1930 our ideas concerning stellar evolution were largely speculative, because nothing definite was known about the sources of stellar and solar energy. It was thought that a young star was also a large and relatively cool one—a vast, super-inflated bubble of highly rarefied gas. Its subsequent life history was envisaged as a process of steady and continuous contraction, with increasing temperature and density, until a point was reached when its matter became too compressed to behave as a gas any longer. Thereafter, contraction would be accompanied by falling temperature. In the advanced stages of senile decrepitude the star would be a low-temperature dwarf; its size that of a planet but its mass that of a star. Finally it would cease to be incandescent and would ultimately cool altogether. As a star it would be 'dead'. Thus in infancy a star would be large and cool, at maturity medium-sized and very hot, and in old age small and cool.

It was not until the decade before the war that sufficiently certain knowledge was obtained as to how and why a star radiates heat and light for our ideas regarding stellar evolution to advance from the speculative to the reasonably assured.

The constant stream of energy that the sun is pouring into space in all directions is costing it over 4 million tons every second. Were it simply 'burning', like a piece of coal, it would be a dead cinder within 5,000 years; yet we know that it has been in existence for something like 3,000 million years. Where, then, does this energy come from?

There is only one possible answer: from within the constituent atoms themselves. Rutherford had disintegrated an atomic nucleus in 1919, and had shown that such a disintegration liberates a stupendously greater amount of energy than any ordinary chemical reaction, which involves only the outer electrons of an atom and leaves the nucleus unchanged. Thus the nuclear reaction that produces helium from a mixture of hydrogen and lithium releases 10 million times as much energy as that resulting from the burning of an equal weight of coal.

Since 1929 our knowledge of reactions involving atomic nuclei has progressed rapidly—witness, unfortunately, the atom bomb. In 1938 Bethe and Weizsäcker independently hit upon the chain of nuclear reactions which would liberate energy at the same rate as the sun. These reactions are impossible to promote at ordinary laboratory temperatures, but are feasible at the high temperatures found in stars; they are for this reason known as thermonuclear reactions.

The picture of a typical stellar life history which developed from this new work was as follows. A young star is large and cool, too cool for any thermonuclear reaction to take place. It would contract under its own gravitation, and in so doing its temperature would rise, until the thermonuclear reaction which occurs at the lowest temperature sets in. As long as there were any materials left for this reaction, the star would remain stable. Then contraction and rising temperature would begin again, until a temperature had been reached at which the next thermonuclear reaction is possible. And so on in a succession of steps until the last hydrogen nuclei had been converted to helium. The star's life is then virtually over; the temperature will fall rapidly and its final contraction, ending in death, will set in.

The white dwarfs are envisaged as being in this final stage. The sun, on the other hand, is still in its prime of life; its temperature will increase until it is about 100 times its present value, and it has an expectation of some 100,000 million years before its final relapse to extinction.

Chapter 8

THE GALAXY, STAR CLUSTERS AND NEBULAE

Is Space Full of Stars?

Every increase in telescopic power shows fainter and fainter stars, banking up in apparently endless perspectives of ever greater remoteness. Nevertheless, the stars do not go on for ever. On the contrary, all the stars that we can see with the naked eye or with the most powerful telescope belong to a single, finite system, or—if we like to regard it as such—gigantic star cluster beyond whose limits lies 'empty space'.

The Milky Way

To the naked eye the Milky Way appears as a luminous band with ill-defined edges, which encircles the entire star sphere. Telescopic study—and still more, telescopic photography—shows that it consists of myriads of faint stars which are crowded so closely together that their integrated light appears as a faint luminosity. This apparent crowding together of stars in the Milky Way is only partly due to real crowding in space, and is primarily dependent on the fact that in the plane of the Milky Way the thickness of stars—or the distance to the edge of the system—is greater than in the direction at right angles to it. In other words the Galaxy of stars is a flattened structure whose major axis is many times longer than its minor axis. Stand a dinner plate on a table and place another plate upside down upon it, and you will gain a fair idea of what the stellar system would look like from outer space.

The Size of the Galaxy

All the figures here given are necessarily approximate and liable to revision, though probably not drastic revision. The major axis, or diameter, of the 'plate' is some 80,000 light years, and the sun is situated near the central plane (since the Milky Way divides the star sphere into two nearly equal portions), about 25,000 light years from the centre of the system; the minor axis or thickness of the Galaxy is about 13,000 light years. All the stars visible to the naked eye lie within 3,000 light years of the sun, showing how small a section of the star system is visible without instrumental aid.

The mass of the Galaxy has been variously estimated as from 30 to 300 thousand million times that of an average star, such as the sun. The whole of this stupendous conglomeration of stars is in rotation about its minor axis, one revolution being completed by the sun in something like 225 million years.

Nebulae and Clusters

The careful searcher of the skies, especially if he is armed with binoculars or a small telescope, will notice many faint and ill-defined patches of light resembling flecks of luminous mist, or minute detached portions of the Milky Way. These may either be vast and distant clouds of glowing gas, known as nebulae, or else clusters of stars whose components are too faint to be distinguished individually. A few of these clusters are near enough to the sun for their individual stars to be well separated and their true nature discerned with quite small instruments or even with the naked eye.

The Galactic or Open Clusters

The Pleiades is one of the brightest and angularly largest of this type of cluster, most of which are invisible without instrumental aid. They are irregular in shape and usually contain several hundred or thousand comparatively bright stars, well separated from one another. They are therefore well suited for observation with small instruments, and many examples are included in Appendix 3.

As their name indicates, they occur most commonly in or near the Milky Way. The brighter clusters are from 100 to 1,000 light years distant, but the majority are between 1,000 and 15,000 light years.

Globular Clusters

A typical globular cluster contains many thousands of faint stars which are condensed so strongly towards the centre of the cluster that even in large telescopes their individual images fuse into an undifferentiated blur of light. They are spherical in shape, great globes filled with stars, and in large instruments are among the most spectacular and gorgeous objects in the heavens.

The globular clusters are all very remote and therefore faint (the brightest are only just visible to the naked eye); the nearest is some 8,000 light years distant and the furthest about 25 times as remote as this. They all appear to be of comparable size, having diameters in the range 25 to 100 light years, though the highly condensed central region is usually not more than five light years across.

Galactic or Gaseous Nebulae

Diffuse, irregular, structureless wisps or clouds of shining gas, occurring chiefly in the Milky Way region. Every conceivable shape, form and size is exhibited by these objects. They are comparatively near to the sun, their distances normally being measured in hundreds of light years. It is improbable that they are self-luminous, and were it not for the high temperature stars that are always found involved in them they would be invisible. Their density is inconceivably low: that of the great Orion nebula, visible to the naked eye and one of the grandest examples of its class, probably does not exceed one-millionth of the most perfect laboratory-produced vacuum.

Planetary Nebulae

In appearance totally unlike the diffuse galactic nebulae, they are so called because their telescopic appearance is reminiscent of that of a planet. They are small, circular, well defined discs of nebulous light, and high magnifications show that there is a high temperature star (30,000°–50,000°) at their centre. Though they may look like flat discs they are actually globes of rarefied gas. About 130 are known; they are mostly faint and unsuitable for observation with binoculars. The distance of the nearest is about 300 light years. The lines of oxygen, hydrogen and helium have been identified in their spectra.

Dark Nebulae

These are clouds of gas, identical with the bright diffuse nebulae but for the fact that they possess no involved stars bright and hot enough to illuminate them. They are therefore dark instead of bright, and can only be seen when silhouetted against the bright background of the Milky Way. They were at one time thought to be 'holes' or starless lanes running through the Galaxy but it has now been established that they are obscuring masses.

The Extragalactic Nebulae

All the objects so far described belong to the stellar system of which the sun is a member. But although space beyond the confines of the Galaxy is largely 'empty' it nevertheless contains many millions of nebulae at distances ranging from under a million to about 1,000 million light years—the latter being the furthest that the Mt. Palomar telescope can reach. Despite the vast numbers of extragalactic objects that are known to exist, space remains to all intents and purposes empty—frighteningly so, perhaps—and they are on the average one million light years apart.

It appears that some of the extragalactic nebulae are gaseous like the galactic nebulae, though of course incomparably larger. Such nebulae are spherical or elliptical in shape and appear to be quite structureless. Others, however, are resolvable (at any rate partially) into separate stellar points by long-exposure photography in conjunction with large telescopes, and the exemplars of this type also have a characteristic spiral structure, somewhat like that of a Catherine wheel. This fact of the stellar resolution of some of the spirals, together with their sizes (deduced from their apparent sizes and their known distances), indicates that they are galaxies of stars similar to our own: island universes. An observer in one of these would describe our Galaxy as an 'extragalactic nebula'. This conclusion is borne out by the determinations of the masses of some of the spirals, the results obtained varying little from 10,000 million times that of the sun.

Not the least intriguing feature of the island universes is the fact that they appear to be scattering helter-skelter in all directions, like fragments of some celestial bomb. The more distant an extragalactic nebula, the greater its velocity of recession. One of the most distant that has yet been investigated is hurtling away from us at a speed of 26,000 miles per second! An alternative to the 'bomb fragment' explanation of this universal recession of the extragalactic nebulae (including the Galaxy) supposes that they are not so much flying *through* space as being carried apart by the expansion of space itself.

As might be expected, considering their vast distances, the extragalactic nebulae are faint objects, only one being visible to the naked eye. This is the great spiral nebula in Andromeda (see p. 138), which owes its relative brightness and apparent size to its nearness, for it is only about one million light years from the sun.

Cosmic Radiation

Since 1902 it has been known that the earth is being continuously drenched in super-short-wave radiation, to which the name 'cosmic radiation' has been given. The shorter the wavelength of a radiation, the more penetrating it is: hence X-rays will penetrate the human body whilst light, except superficially, will not. X-rays, however, are stopped by a few millimetres of lead, and protection against even the γ-rays emitted from radio-active materials is given by a few inches of lead. Yet cosmic radiation can penetrate a block of lead to a depth of 16 feet. It has been identified in deep mines and at the bottoms of lakes, as well as high in the atmosphere. Since it is stronger in the upper atmosphere than at sea level it clearly comes from outside the earth; and since it is equally strong by day and by night it presumably does not originate in the sun.

The source, and even the nature of this extraordinary radiation is still a mystery.

In the early 1930's Jansky found that radiation of radio frequency is coming in from somewhere beyond the Solar System. Later work with modified radar equipment and a special 'radio telescope' (a 31-foot metal mirror with a dipole aerial at its focus)[1] has revealed that this long-wave radiation is received only from a belt centred upon the Milky Way, and not more than about 60° wide. Furthermore, its intensity is greatest in the direction of the galactic centre (Sagittarius) and weakest in the opposite direction.

Since the pioneering experiments of Jansky, the field of radio astronomy has advanced to become at least as important as optical astronomy. However, any discussion of the fascinating discoveries and mysteries revealed by this exciting observing technique would be far beyond the scope of this book.

A Scale Model

It will be interesting to see how the stars and extragalactic nebulae fit into the scale model of the Solar System which we constructed in Chapter 4. There, it will be remembered, we reduced all the dimensions of the Solar System about 30 million times, so that in the model the sun was a globe 150 feet in diameter, sitting in Trafalgar Square, and Pluto was out at Yeovil. But even this scale is too large to include the stars conveniently, since Proxima Centauri would be over 800,000 miles away—more than three times the actual distance of the moon.

We will therefore shrink the model still further, until the whole Solar System, bounded by Pluto's orbit, is no larger than a halfpenny, lying at the foot of Nelson's Column. On this scale, the nearest star will be found on the steps of the National Gallery on the north side of Trafalgar Square—though to say 'found' is misleading, because it would be far too small to be seen. The Orion nebula and the Pleiades will be situated respectively in Richmond Park and Clapham Common.

[1] The Jodrell Bank radio telescope is 250 feet in diameter.

The galactic centre will be nearly 400 miles from London—at Dijon, say. The Galaxy will then include England and Wales, France and the Low Countries, Germany, Switzerland and Austria, and about half of Spain, Italy, Czechoslovakia and Denmark.

The Andromeda spiral, one of the nearest extragalactic nebulae and the only one visible to the naked eye, will be 12,000 miles from London. The remotest objects within the reach of the Mt. Palomar telescope will be 4 million miles away.

Our universe is indeed an agoraphobe's nightmare.

Chapter 9

ASTRONOMICAL INSTRUMENTS

Little can be learned of the nature of the heavenly bodies, or of the structure of the universe beyond the frontiers of the Solar System, without the help of instruments. Indeed, the story of man's widening knowledge of the physical universe during the last three and a half centuries, which comprise the modern period in astronomy, is closely paralleled by the story of his invention and refinement of a whole series of instruments and the techniques of their use.

Roughly, this period can be divided into three stages on the basis of instrumental development. First, the telescope was the great fact-finder; the second stage was ushered in by Fraunhofer and Kirchhoff when they founded the science of celestial spectroscopy during the mid-nineteenth century; third, the camera replaced the observer's eye. Today the telescope, the spectroscope and the camera are still the astronomer's most powerful instruments of research. It is possible that we are now on the threshold of a fourth great period of discovery, based upon the application of radio techniques to astronomy.[1]

The Refractor

The modern period in astronomy can conveniently be dated from 1609, when Galileo for the first time turned the recently invented telescope towards the heavens.

In its most primitive form the refracting telescope consists of two convex lenses, which are looked through in line; one is of long focal length and the other of very much shorter, the latter being the lens nearer the observer's eye.

[1] See pp. 46, 66, 81, 98.

A convex lens has the power of bending or refracting to a focus the light that passes through it. Figure 10 (a) shows how a lens of this type forms an image of a point source such as a star, which is so distant that its rays may be regarded as parallel. This is a real image, and could be caught on a screen if needs be, just as the sun's image is caught on a piece of paper when a convex lens is used as a burning glass.

In the telescope this real image formed by the first lens, or object glass, is not viewed by the naked eye, but is itself magnified by the

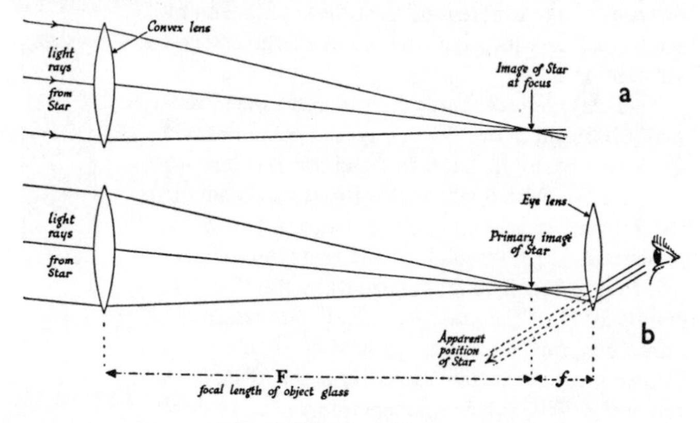

FIGURE 10

second lens (the eyepiece); this acts exactly like a magnifying glass, although it is not magnifying a physical object, but the image of an object formed by the object glass (see Figure 10 (b)). In practice, the eyepiece usually consists of a pair of lenses, but for the sake of simplicity we shall speak only of single-lens oculars.

Such a telescope is far from perfect, because the simple convex lens fails to bring all the rays from a point source accurately to a point focus. The cause of most of the trouble is called chromatic aberration, which seriously impeded the development of the refractor during the first 160 years of its existence. Briefly, the lens has some of the properties of a weak prism: it does not simply refract the light rays to a focus, but refracts them by amounts which vary with their wavelength (or colour), thus converging them to different foci. Blue light is brought to a shorter focus than red, with the result that no matter how we adjust the focusing rack of the eyepiece, nowhere will a sharply defined image be found.

It was discovered early in the seventeenth century that the effects of chromatic aberration could be ameliorated by enormously increasing the focal length of the object glass as compared with its diameter. Unwieldy great telescopes, sometimes up to 200 feet long, were constructed. Slung from poles, impossible to direct at will to different parts of the sky, shaken by every gust of wind, their practical usefulness was virtually nil.

Two ways of escape from this impasse were eventually discovered. In 1758 an English optician named Dollond constructed the first achromatic lens—a lens, that is, which is reasonably free from chromatic aberration.[1] It is a compound lens, the two components of which are made of different kinds of glass (see Figure 11). By choosing carefully the chemical composition of the glass it is possible to retain the refracting, image-forming properties of the combination, while at the same time the dispersive properties of the two components more or less cancel one another out.

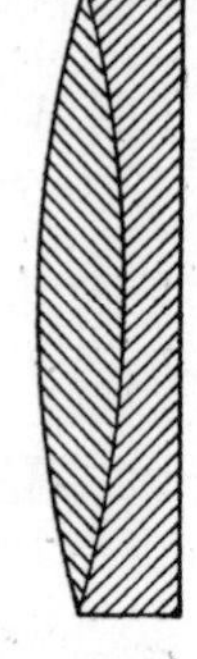

FIGURE 11

With this development the refractor took on a new lease of life, and during the next hundred years larger and larger instruments were built. This phase culminated in 1895 with the Yerkes refractor, whose object glass measures 40 inches across. But two new impediments in the way of the further development of the refractor had in the meantime been encountered; nor have they yet been circumvented—indeed, they appear to be insuperable. In the first place it is difficult to the point of impossibility to cast glass discs for lenses of this size which are optically homogeneous throughout. Secondly, and even more disastrous, since the object glass can

[1] Chester Hall, a scientifically-minded gentleman of leisure, had discovered the principle of the achromatic lens twenty-five years previously, but his discovery had attracted no attention and was quickly forgotten.

of necessity be supported at its edge only, the glass begins to sag under its own weight when the diameter exceeds about 4 feet. The distortion is of course very slight, but it is nevertheless sufficient to produce a perceptible deterioration in the quality of the image.

For these reasons, and also because of its imperfect achromatism, the refractor has today taken second place to the reflecting telescope, and all the large modern instruments are reflectors.

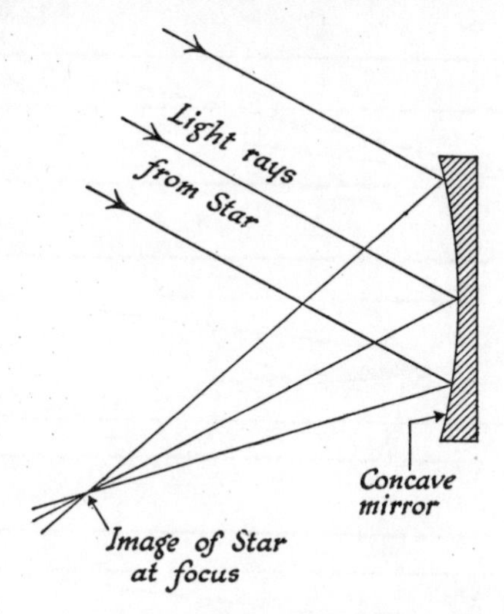

FIGURE 12

The Reflecting Telescope

The principle of a telescope based not upon the refraction of light by a lens but upon its reflection by a mirror was worked out by James Gregory in 1663, the first telescope of this type being constructed by Newton five years later. His aim was to bypass chromatic aberration, which at that time had brought the development of the refractor to a standstill.

As in a refractor, the reflecting telescope has two main components: an objective which gathers the light to a focus, and an eyepiece which magnifies the image so formed. But the objective is a concave mirror, not a lens. Comparison of Figures 12 and 10 (a) will show at a glance that a concave mirror is just as capable of forming an image as a convex lens. More so, indeed, for light of all wavelengths is brought to a focus at the same distance from the objective; the mirror is absolutely free from chromatic aberration.

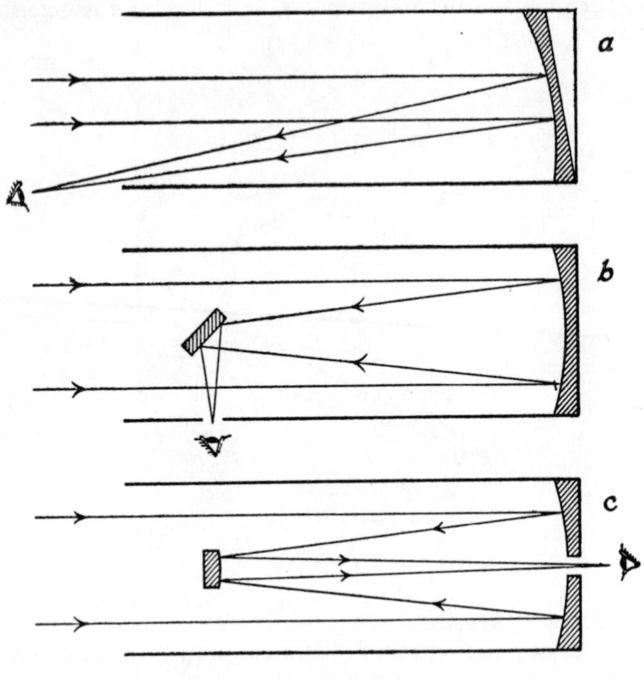

FIGURE 13

In order to bring rays reflected from all parts of the mirror to the same focus, its curve has to be very carefully figured. A spherical mirror is found to suffer from the defect that it reflects rays from its outer regions to a shorter focus than rays from near its centre. To correct this, the centre has to be ground deeper until its curve is no longer a section of a sphere, but a parabola.

In actual construction the reflector permits of greater variation than the refractor. The practical problem is how to get the eyepiece and the observer's eye behind the image without his head getting in the way of the light passing down the tube to the mirror. Figure 13 shows three solutions of this problem.

In (a), known as the Herschelian reflector, the mirror is tilted slightly, so as to throw the reflected cone of rays to the side of the tube where they are received by an observer sitting with his back towards the object he is looking at. This form is now obsolete, the commonest today being that shown in (b), the Newtonian reflector. Here the reflected cone is intercepted by a small plane mirror suspended centrally inside the tube near to the focus, and thence turned through a right angle to an eyepiece mounted in the tube; here the observer sits sideways-on to the object he is looking at.

The Cassegrain reflector, illustrated diagramatically in (c), is also much used at the present day, having several advantages over the Newtonian; one of these is the very much longer focal length it permits in a tube of the same size—the significance of which will appear when we come to discuss telescopic magnification. In the Cassegrain the flat secondary mirror of the Newtonian is replaced by a small convex mirror, mounted axially, which reflects the converging rays back down the tube and through a hole cut centrally in the main mirror; the eyepiece is mounted axially behind this, the observer facing the object he is observing, as with a refractor.

Large reflectors often combine the Newtonian and Cassegrain arrangements in the same mounting. The mirror is bored centrally, and eyepiece positions are built behind the main mirror and also at the top of the tube. Used as a Newtonian, the secondary mirror is a flat, swung into position near the mouth of the tube; used as a Cassegrain, this is replaced by an axially mounted convex secondary mirror.

The development of the reflector, like that of the refractor as well as that of true love, did not run smoothly. At first the mirrors were ground from discs of a highly reflecting alloy known as speculum metal. This tarnished rather quickly, however, and repolishing involved some degree of regrinding—a lengthy and exacting process, particularly in the case of large mirrors. The achromatic objective having in the meantime been invented, the limelight switched back to the refractor, and reflectors remained out of favour until the middle of last century, when a chemical method of depositing a very thin and highly reflecting layer of silver on glass surfaces was discovered. When tarnished, this could be wiped off with a wet rag and a new coating deposited. Glass immediately replaced speculum metal as the material for telescope mirrors, and has remained so until the present day. Two comparatively modern refinements are the use of pyrex instead of ordinary glass, since it is far less affected by temperature changes; and a method of depositing vaporized aluminium on a glass surface in a vacuum chamber, the aluminium film being tougher and more resistant to tarnish than silver.

Since the mirror can be supported from the back, as well as at the sides, flexure can be completely eliminated; and since no light passes through the mirror, but is reflected from its surface, it is unnecessary for the glass to be optically perfect.

What the Telescope Does

Nine people out of ten, asked what a telescope is for, will reply: 'To make things appear larger,' 'To make distant objects seem nearer,' or something to the same effect—i.e. to magnify. But magnification is only one, and not the most important, of the telescope's functions.

(a) *Magnification.* The magnifying power of a telescope depends solely upon the relative focal lengths of the objective and the eyepiece. It is independent of everything else, including the size of the objective. If the focal length of the objective (F, in Figure 10 (b)) is 60 in., and that of the eyepiece (f) is 2 in., then the telescope will magnify thirty times. If we wish to double the magnification we simply change the eyepiece for one whose focal length is half that of the first, namely 1 in. Thus it is quite mistaken to suppose, for example, that the Mt. Palomar telescope (the diameter of whose mirror is 200 in.) is more powerful than the Mt. Wilson 100-in. reflector in the matter of magnification; in either case, if the focal length of the ocular is 1/100th that of the objective, the magnification will be ×100.

It might be deduced from this that merely by using objectives of longer and longer focal length, and oculars of the shortest focal length practicable, there would be no limit to the magnifications that could be obtained telescopically—that, for example, a magnification of ×12,800 could be obtained at the Cassegrain focus of the Palomar instrument (which is 266 ft. 8 in. from the main mirror) by inserting an eyepiece of ¼-in. focal length in the drawtube.

The image—what could be seen of it—would then admittedly be nearly 13,000 times larger than the object as seen with the naked eye, but its quality would be so inferior that less detail would be discernible than in a very much less highly magnified image. The reason for this is that astronomers do not work in a vacuum: between them and the objects at which they point their telescopes are several hundred miles of miscellaneous gases, at different temperatures, of different densities, and all in movement —the earth's atmosphere. Thermal currents in the atmosphere cause the same sort of deterioration of the telescopic image as can be observed when looking through the hot gases above a bonfire or chimney: the object thus seen appears to be shimmering or 'boiling', and all its finer detail is lost. Trouble from this cause increases proportionally with the magnification. It is for this reason also that the sites of modern observatories are chosen with great care, for the general quality of the 'seeing' varies considerably in different localities.

It is found in practice that on nights when the seeing is good, magnifications up to ×60 or ×100 per inch of aperture are practicable with small and moderate-sized instruments. But with large instruments this proportion falls off rapidly, and no telescope in existence, even the largest, can profitably employ magnifications of more than about ×3,000, and then only during the infrequent intervals of superlatively good seeing.

(b) *Light grasp.* Astronomical objects are not only inconspicuous or invisible because of their smallness, but also on account of their faintness. And obviously it is no use magnifying an object that is too faint to be seen anyway.

The more light that enters the eye from an object such as a star, the brighter that object will appear. It is for this reason that the pupils of our eyes dilate in the dark. In the case of a telescope, the image of a star is formed of all the light that has fallen on the objective, and thence been reflected or refracted to the focus. If the area of the objective is 100 times that of the pupil of the naked eye, the star will appear 100 times brighter when observed through the telescope than when seen without it. To take an actual example: the area of the pupil of the dark-adapted human eye averages 1/9 square inch; the area of the reflecting surface of the Mt. Palomar mirror is a little more than 209 square feet; hence the light grasp of the great reflector exceeds that of the naked eye by a factor of about 270,000. Since the most distant astronomical objects are, generally speaking, the faintest, it can be seen what an immensely powerful instrument this is for probing the hitherto unexplored depths of space.

(c) *Resolution.* The third function of the telescope is likewise dependent upon the linear size of the objective which forms the image. The objective's resolving power determines the size of the finest detail that the image contains, and therefore the finest detail that will be visible no matter how far the magnification may be pressed; alternatively, it determines how widely separated a pair of stars must be before the telescope will show them as individual stellar points, using the maximum useful magnification.

The telescopic image may be likened to a half-tone newspaper photograph. The figure of a man, 2 in. high, will be more detailed in a fine-screen reproduction than in a coarse-screen one. Some details may be seen more clearly in either photograph by looking at it with a magnifying glass. But no amount of magnification can reveal detail which is smaller than the grain of the photograph, since beyond a certain point magnification breaks up the photograph into a pattern of the separate dots which compose it.

So with the telescopic image. The fineness of detail in the primary image is determined by the size of the objective. Magnification of this image by the ocular is useful up to the point at which the finest detail in the image is large enough to be seen. Further magnification will make these details larger but will reveal no new ones.

In order to bring still finer detail into visibility, a larger objective must be used.

Thus, while the limit of resolution of a 1-in. telescope is 4·5″, that of the Mt. Palomar telescope is 0·02″—the angle subtended by a halfpenny at a distance of something like 150 miles.

Summary

The function of the astronomical telescope is therefore threefold: to magnify small objects; to render faint objects bright enough to be seen and studied; and to bring to light minute detail, and separate closely adjacent objects, whether details of a planetary surface or the components of a double star.

Telescope and Spectroscope

The telescope's second function, that of gathering light and intensifying it in the image, is made use of to the exclusion of the others whenever it is used in conjunction with a spectroscope. Its sole use then is to collect and feed into the subsidiary instrument enough light for the latter to produce an observable image.

Without the telescope we should have learnt nothing of the spatial arrangements of the universe beyond the Solar System; without the spectroscope we should have learnt nothing of its nature. The spectroscope, though dependent on the telescope for providing it with sufficient light from the object under observation, is no less powerful and valuable a research instrument in its own right.

Astronomical Spectroscopes

By passing light through a prism (or reflecting it from a diffraction grating consisting of a plane surface graved with very numerous, very fine parallel lines) it can be split up into its constituent wavelengths and colours. Light is refracted on passing through a prism, just as through a lens, but all wavelengths are not diverted to the same extent. The shorter the wavelength the larger the angle through which the light is refracted. Hence the action of a prism is to sort out mixed radiation like sunlight into its component colours (Figure 14). If a narrow slit, cut in a screen and illuminated by sunlight, is looked at through a prism, it will be found that its image is drawn out into a band of colour, known as a spectrum; this will be red at one end, merging into orange, yel-

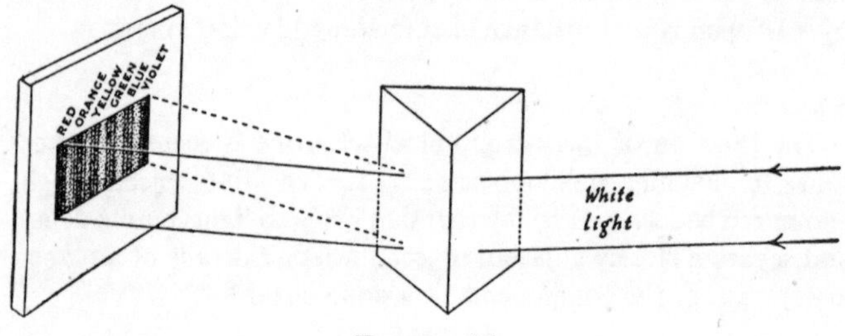

FIGURE 14

low, green and blue, to violet at the other end. If the sunlight is shut out, and the light from a sodium flame substituted, it will be found that the continuous spectrum has been replaced by a single pair of bright yellow lines—the images of the slit in the only two wavelengths which comprise the radiation of incandescent sodium.

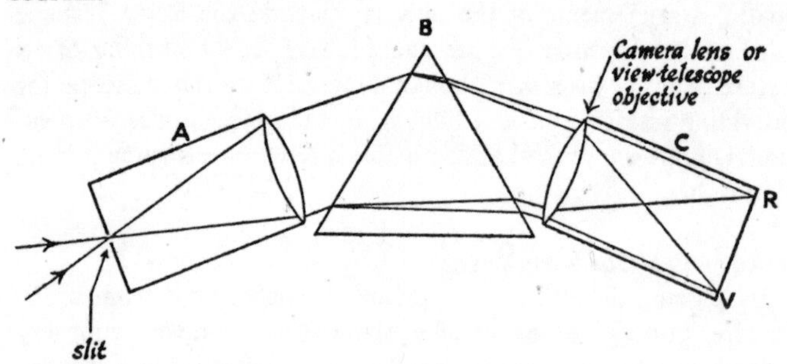

FIGURE 15

A variety of spectroscope designs, both of the diffraction grating and prism types, have been developed for astronomical purposes. The basic prismatic type consists of the three components shown in Figure 15: *A*, the collimator; *B*, the prism; *C*, the camera or viewing telescope.

The collimator consists of a very fine slit which lies in the focal plane of the telescope's objective, so that the primary image of the star falls upon it. The collimating lens lies behind it at a distance equal to its own focal length; it therefore converts the diverging rays from the slit into a parallel beam. After refraction by the prism, all the rays of a particular wavelength are parallel to one another but diverging from the parallel bundle of every other wavelength. Each of these bundles is then focused to a line-image of the slit in the plane *VR* by the camera lens; in this plane the photographic plate is placed. If the spectrum is to be observed visually, the plate is removed and an eyepiece placed behind the focal plane of the camera lens, which now acts as the object glass of the viewing telescope.

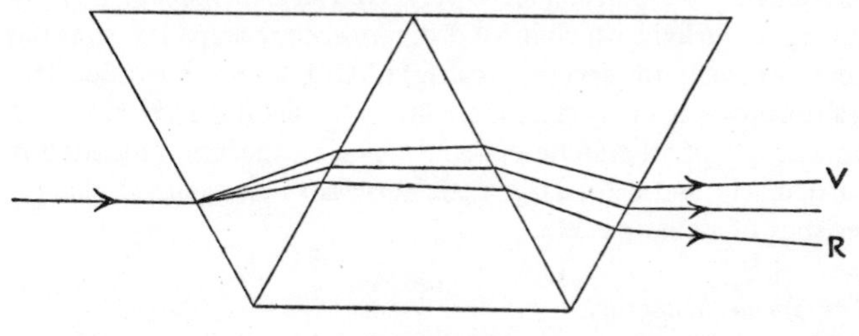

FIGURE 16

In order to obtain greater dispersion—though at the cost of a fainter spectrum—the light may be led through a train of prisms in succession, each prism increasing both the deviation and the dispersion of the rays.

A simple form of spectroscope for visual observation is that employing a direct-vision prism. This consists of an odd number of prisms of alternate crown and flint glass (the same two varieties of glass used in the achromatic lens) which are cemented together as shown in Figure 16. The dispersive power of flint glass being greater than that of crown, the two crown prisms annul the deviation of the central ray whilst leaving the dispersion—or separation of the red from the violet—unaffected. When used as a stellar spectroscope no slit is needed, the stellar image being so small as itself to constitute a minute length of very narrow slit. The spectrum of a star is a thin, coloured line which can, if necessary, be broadened by a special form of lens to enable its details to be more easily seen.

One of the oldest forms of astronomical spectroscope is the objective prism. This is a small-angled prism (there may be several) mounted in front of the telescope's objective and completely covering it. Again, a slit is not necessary owing to the smallness of the stellar images; nor is a collimator, since the rays from the enormously distant stars are already sensibly parallel. The telescope itself now plays the role of the view telescope in an ordinary spectroscope.

The objective prism, though it has its limitations, is a valuable instrument on two accounts. Owing to its small dispersion and the saving of the light which in an ordinary spectroscope is lost at the jaws of the slit, the spectra are bright. Also, several hundred spectra (or however many stars there are in the field) may be recorded on a single photographic plate at a single exposure, rendering it particularly useful for large-scale surveys of the general characteristics of stellar spectra.

The Spectroheliograph

The most ingenious development of the spectroscope is undoubtedly the spectroheliograph, invented independently by Hale, an American, and Deslandres, a Frenchman, in 1890. Since that date, and with its help, our knowledge of the physical and chemical structure of the sun has advanced at an unprecedented pace.

Figure 17 represents the instrument diagramatically. O_1 is the objective of the telescope to which is attached a high-dispersion

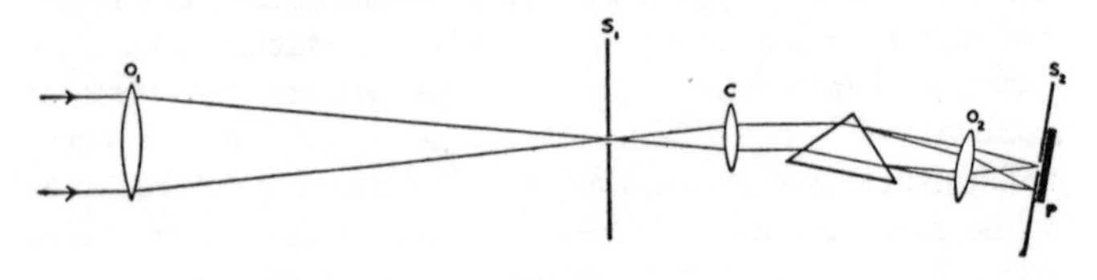

FIGURE 17

spectroscope with its slit, S_1, in the focal plane of O_1. The collimating lens (C), prism or grating, and camera lens or view-telescope objective (O_2) are arranged as in a normal spectroscope; but instead of the focal plane of O_2 (in which the spectrum is brought to a focus) being occupied by the photographic plate, a screen with a second slit (S_2) is mounted there. Immediately behind S_2 lies the photographic plate, P. The whole of the solar spectrum therefore falls on the screen except for the narrow strip which is allowed to pass to the photographic plate through S_2. S_2 is movable transversely (i.e. along the axis of the spectrum) and is adjusted in width until it just admits a single line of the solar spectrum, occluding the continuous spectrum on either side of it.

Suppose that the position of the second slit is adjusted so as to allow the H line of calcium to fall upon it. Both slits being tall enough to cover the whole of the sun's image, the photographic plate will record the distribution of calcium light along the narrow strip of the solar image that is covered by S_1. If the whole spectroheliograph, including the two slits, is now displaced laterally one slit-width, the main telescope and the plate remaining stationary, a second strip of sun in calcium light will be recorded on the plate alongside the first. By covering the whole solar image in this way, its complete photograph in the light of any selected wavelength can be built up. The discovery of the calcium and hydrogen flocculi and the hydrogen 'flares' by this method has been described in an earlier chapter.

Astronomical Photography

It has been said that the photographic plate is second only to the telescope in importance, so far as the astronomer is concerned. Since the earliest daguerrotypes of the sun and moon were made a century ago, the development of photographic technique and materials has to a large extent revolutionized the practice of astronomy.

In two particular respects photographic is superior to visual observation: as a time-saver, and as a means of investigating, by lengthy exposures, objects which are too faint to be satisfactorily studied visually. The former characteristic is exemplified in the construction of star charts, in the discovery of comets, asteroids, variable stars and novae, and by the fact that plates can be studied and measured during the daytime, thus liberating the telescope for other work during the precious night hours. Its power as an instrument for the observation of very faint objects is indicated by the fact that the study of the extragalactic nebulae is today made entirely by photography.

It is markedly inferior to visual observation wherever resolution is the paramount factor, since the resolving power of the human eye is greatly superior to that of the photographic plate, especially when the latter has to be exposed for long periods. Thus the lunar photographs taken with the 100-inch reflector, superb though they are, reveal no detail that cannot be seen with telescopes one-tenth the size.

Telescopes and Cameras

The achromatic objective largely, but not entirely, eliminates chromatic aberration. What it does is to bring the green and yellow rays (to which the eye is most sensitive) to one focus, and the visually dim red and violet rays to another, rather farther away. The photographic plate, however, differs from the eye in being most sensitive to the blue rays, and in order to reduce exposures it is therefore these that should be brought to a focus at the surface of the plate. Objectives corrected in this way are called photographic objectives, and are useless for visual observation. With the plate mounted at the primary focus, the entire telescope becomes in effect a gigantic camera.

As a photographic telescope, the reflector is superior to the refractor, owing to its perfect achromatism: its visual and photographic foci coincide, so that the image on the plate is accurately focused in all colours and not only in those to which the emulsion is most sensitive.

Primary focus photography has the disadvantage that the field of good definition is small. More than a few degrees from the axis of the telescope, aberrations which distort the image come into play. For wide-field photography, special photographic lenses—consisting of three, four or more components—have to be used.

The camera in this case is attached to the telescope, which is used only for 'guiding'—that is, for following the star in its diurnal motion so that its image remains in exactly the same position on the plate throughout the exposure. Owing to the necessity of taking long exposures of continuously moving objects, celestial photography would be impossibly difficult were it not for the equatorial form of telescope mounting. In this, the telescope is capable of being rotated about two mutually perpendicular axes, one of which, the polar axis, is set up parallel to the earth's axis—pointing, that is, at the celestial pole. If, now, the telescope is directed at the required star and the polar axis made to turn at the rate of one rotation every 23 hr. 56 min.,[1] the star will remain in the centre of the field indefinitely. But even though the rotation is imparted to the polar axis by the most perfect mechanism possible, small remaining inaccuracies and the fluctuating conditions of the atmosphere make it always necessary for an observer to 'guide' the telescope throughout an exposure.

The Schmidt Camera

The 'faster' the optical system of his camera, the fainter will be the objects that the astronomer can record in a given exposure time, and the shorter will be the exposure necessary to record a given object satisfactorily.

The 'speed' of a camera lens, as far as extended images are concerned,[2] is determined by its focal ratio—its focal length divided by its aperture. A 10-in. lens of 63 in. focal length will require exposures of the same length as a lens whose aperture is only 5 in. and focal length 31½ in., since the focal ratio is the same in each case: f/6·3.

But as the focal ratio is reduced, so the field of good definition gets smaller and smaller, the reason being that off-axis aberrations (affecting the quality of images lying far from the axis) are exaggerated when the aperture of an objective is increased at the expense of its focal length.

The designer of astronomical cameras was therefore in a dilem-

[1] See p. 24.

[2] In the case of point images, 'speed' is a function of aperture only.

ma: one horn represented by 'fast' cameras with very small usable fields, the other by cameras giving good definition over wide fields but requiring very long exposures.

This dilemma was brilliantly resolved by Schmidt in 1930. We have seen that the spherical aberration which ruins the performance of a spherical mirror can be removed by slightly deepening its centre, converting it from a spheroid into a paraboloid. The price of this freedom from spherical aberration is the introduction

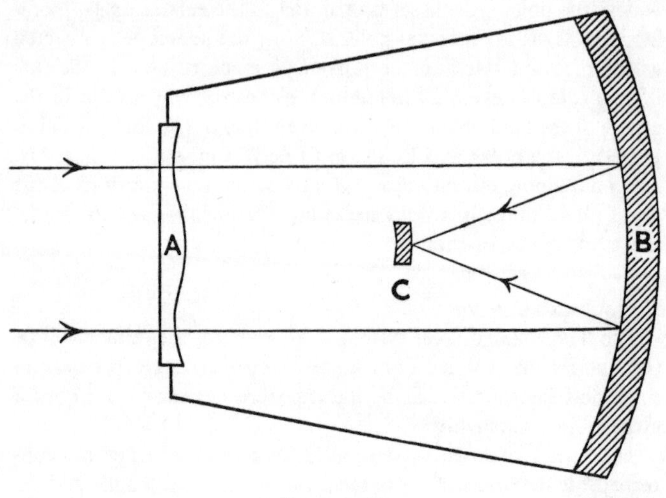

FIGURE 18

of those very aberrations mentioned above; for visual purposes this is not of great importance since very wide fields are not normally required. Schmidt showed how it is possible to eliminate the aberration of a spherical mirror without reintroducing the off-axis aberrations of the paraboloid in exchange, thus opening the way for wide-field systems of exceptional 'speed'.

The Schmidt camera has been developed into a large variety of forms; Figure 18 represents the prototype design. *B* is the spherical mirror. *A*, the correcting plate, is the crux of the thing: the convex and concave curves of its surface are exactly those required to distort the beam passing through it in a manner equal but opposite to the spherical aberration of the mirror. Hence the two cancel one another out, and the images formed in the mirror's focal plane at *C* are perfect over a field many times wider than in any other optical system of comparable photographic speed.

An ordinary parabolic mirror, with a focal ratio of f/1, would have a well defined field measured in ″ arc; that of the f/5 Mt. Wilson reflector is about 7′ in diameter, and the Mt. Palomar instrument's only about half this. Yet an f/1 Schmidt can cover a field of something like 20° diameter, and record in a few minutes a nebula so faint that an ordinary wide-angle photographic lens would require several hours' exposure to reveal it.

APPENDICES

SUMMARY OF DATA RELATING T

(1)	(2)	(3)	(4)	(5)	(6)	(7)
	Symbol	*Mean Solar Distance (miles)*	*Period of Revolution*		*Maximum Distance from Ecliptic*	*Angular Equatorial Diameter*
			(days)	*(years)*		
Mercury	☿	35,950,000	87·97	0·241	7° 0′ 14″	4″·7–12″·9
Venus	♀	67,180,000	224·70	0·615	3° 23 39″	9″·9–64″·0
Earth	⊕	92,870,000	365·26	1·000	—	—
Mars	♂	141,500,000	686·98	1·881	1° 51′ 0″	3″·5 –25″·
Ceres	①	257,200,000	1,681·45	4·603	10° 36′ 56″	0″·27–0″·6
Jupiter	♃	483,300,000	4,332·59	11·862	1° 18′ 21″	30″·5–49″·8
Saturn	♄	886,100,000	10,759·20	29·458	2° 29′ 25″	14″·7 –20″·
Uranus	♅	1,783,000,000	30,685·93	84·015	0° 46′ 23″	3″·4 – 4″·2
Neptune	♆	2,793,000,000	60,187·64	164·788	1° 46′ 28″	2″·2 – 2″·
Pluto	♇	3,666,000,000	90,600	248.4	17° 8′ 34″	0″.23
Sun	☉	—	—	—	—	31′ 59″ (mea
Moon	☾	238,857 (from ⊕)	27d. 7h. 43m. 11·5s.		5° 8′	31′ 5″ (mean

Notes

Column (6): Gives the inclination of each planet's orbit to the ecliptic.

Column (7): Explains why no planet save Jupiter, and Venus in the crescent phase, presents a sensible disc to small instruments such as field glasses. 1″ is the angle subtended by a halfpenny at a distance of 3½ miles.

Column (12): These figures depend upon the mass of the planet and also upon its radius.

Column (13): Gives the apparent brightness of the sun, moon, and

THE SUN, MOON, AND PLANETS

(8)	(9)	(10)	(11)	(12)	(13)	(14)
Linear Diameter (miles)	*Mass* (⊕=1)	*Volume* (⊕=1)	*Period of Axial Rotation*	*Surface Gravity* (⊕=1)	*Stellar Magnitude*	*No. of Moons*
3,025	0·04	0·06	59.4d.	0·26	−1·2 to +1·1	0
7,600	0·83	0·88	244d.	0·90	−4·3 to −3·3	0
7,927	1·00	1·00	23h. 56m. 4.1s.	1·00	—	1
4,200	0·11	0·15	24h. 37m. 22.7s.	0·38	−2·8 to +1·6	2
490	? 0·000125	0·0002	9h. 05m.	?0·04	7·1	—
88,700	318·4	1,312	9h. 50m.–9h. 55m.	2.66	−2·5 to −1·4	12
75,100	95·2	763	10h. 14m.–10h. 39m.	1·14	−0·4 to +1·2	10
30,900	14·6	59	10h. 45m.	0·96	5·7	5
33,000	17·3	72	?15h. 40m.	1·00	7·6	2
<4,250	*c.* 0·1	*c.* 0·09	6.39d.	?	13 to 15	? 0
864,000	333,434	1,300,000	24d. 15h. 36m. (equatorial)	27·89	−26·9	—
2,160	0·012	0·02	27d. 7h. 43m.	0·16	−12·5 (Full)	—

planets expressed in terms of the Stellar Magnitude scale (see p. 86). Roughly speaking, the brightest stars visible to the naked eye are magnitude 1, and the faintest magnitude 6. A star as much brighter than a first magnitude star as the latter is than a second magnitude star would be magnitude 0. Similarly, a star as much brighter than a magnitude 0 star as this is brighter than a sixth magnitude star will have a magnitude of −6. And so on. Thus it can be seen from column (13) that of the major planets, only Pluto is invisible with binoculars, although Neptune is invisible to the naked eye and Uranus only just visible.

Appendix 2

TIME AND DIRECTION

Local Time by the Stars

There is an ingenious method of telling the Local Time at night without a watch, the line from Polaris to β Cassiopeiae (the star at the right-hand end of the Cassiopeia 'W') being used as an hour hand.

The procedure is to imagine a 24-hour clock dial superimposed upon the sky with Polaris at its centre. Estimate the dial number nearest the line joining Polaris and β Cassiopeiae and add it to the number of the current month as given in the Table below:

January	16	July	4
February	14	August	2
March	12	September	24
April	10	October	22
May	8	November	20
June	6	December	18

If the sum comes to more than 24, subtract 24. Allowance must of course be made for Summer Time, if in force.

Suppose, for example, that on a December evening β Cassiopeiae is directly above the Pole Star, i.e. the 'hour hand' points to 24. The month number is 18, hence the Local Time is 24 + 18 = 42 hours. Subtracting 24 we get 18 hours, or 6 p.m.

This rough and ready method is only strictly accurate at the middle of each month, but at any time it will give the correct answer within an hour or so.

Finding One's Way by the Stars

Unless clouds obscure the sky no one who has even a nodding acquaintance with the heavens can ever be completely lost at night—lost, in the sense that he does not know whether he is facing north, south, east or west.

The cardinal points can be defined with fair accuracy by eye estimations based upon an observation of Polaris:

(i) If Polaris, easily found by reference to the Plough, is visible, stand facing the point on the horizon which is judged to be vertically below it. North is then in front of you, south behind you, east on your right hand and west on your left.

(ii) The Pointers of the Plough can also be used as a compass. When the Plough is 'right way up' (Figure 1, Position 1), they are pointing south—i.e. if the line through them is continued right over the star sphere through the zenith, it will cut the horizon at the due south point. Similarly, in Position 2 they are pointing west, in Position 3 north, and in Position 4 east.

Should the northern sky be obscured there are supplementary indications to work on:

(iii) Other bright and easily found pointers to the Pole Star, besides the Pointers and the Plough, include:

<table>
<tr><td>Rigel→Capella→N.</td><td rowspan="3">Winter stars</td></tr>
<tr><td>Spica→2nd star from end of Plough's handle→N.</td></tr>
<tr><td>Procyon→midway between Castor and Pollux→N.</td></tr>
<tr><td>Altair→western arm of the Cygnus cross→N.</td><td rowspan="3">Summer stars</td></tr>
<tr><td>Either side of the Great Square of Pegasus→N.</td></tr>
<tr><td>N.E. corner of Gt. Sq. of Pegasus→β Cassiopeiae→ N.</td></tr>
</table>

(iv) Objects on the celestial equator rise due east and set due west. If any of the following stars are observed to have just risen or to be about to set, therefore, they give accurate information upon which it is possible to orient oneself:

	Object	*Distance from Celestial Equator, in degrees*
Aquarius:	α (mag. 3)	1
	the 'Water Jar' (mags. 4)	within 2
Orion:	the Belt stars (mags. 2)	within 2

APPENDIX 2

	Object	Distance from Celestial Equator, in degrees
Cetus:	α (mag. 3)	4
	γ (mag. 3)	3
Procyon =	α Canis Minoris (mag. 1)	6
Head of Hydra including α (mag. 3)		within 7
Virgo:	β (mag. 3)	3
	γ (mag. 3)	1
	ζ (mag. 3)	½
Serpens:	α (mag. 3)	7
	μ (mag. 3)	3
	η (mag. 3)	3
Ophiuchus	δ and ϵ (mags. 3)	within 5
	β (mag. 3)	5
Aquila:	α (mag. 1)	9
	δ (mag. 3)	3
	θ (mag. 3)	1

(v) Stars further from the celestial equator also have their usefulness:

Stars	
Pleiades Castor and Pollux	rise N.E. and set N.W.
Aldebaran, Regulus, Great Square of Pegasus (southern side)	rise E.N.E. and set W.N.W.
Sirius Spica	rise E. to E.S.E. and set W. to W.S.W.
Antares Fomalhaut	rise E.S.E. to S.E., set W.S.W. to S.W.

(vi) Orientation is also possible by many small indications which will not be lost upon the man who is familiar with the constellations. For example:

The cross of Cygnus is N.E. when it is lying on its side near the horizon.

The cross of Cygnus is N.W. when it is upright, near the horizon.

Orion's belt stars are E when they are vertical.

Aldebaran is E. when it and the Pleiades are perpendicular to the horizon.

Altair is E. when it and its two companions are perpendicular to the horizon.

Orion's belt stars are W. when roughly horizontal, near to the horizon.

Aldebaran is W. when it and the Pleiades are horizontal.

The Altair triplet is W. when it is horizontal.

Orion's belt is S. when it makes an angle of 45° with the horizon.

The centre of the Great Square of Pegasus is S. when its bottom side is horizontal.

Procyon is S. when vertically below Castor and Pollux.

(vii) Owing to the obliquity of the ecliptic, the moon and planets are unreliable guides. The following rough rule is useful, however:

At 1st Quarter (7 days old) the moon is S. at about 6 p.m.

At Full (14 days old) the moon is S. at about midnight.

At 3rd Quarter (21 days old) the moon is S. at about 6 a.m. At intermediate phases it will of course be S. at intermediate times. When 10½ days old (half-way from 1st Quarter to Full), for example, it will be due S. at about 9 p.m.

With these indications the cardinal points can be determined at night within 10° easily. For more accurate purposes, tables have been compiled which give the compass bearings of representative bright stars at every hour of the year.

(viii) By day, the sun is, of course, the direction finder:

At rising, the sun is always within 23° of due E.; and at setting, within 23° of due W. Its true bearings throughout the year are summarized below:

Time of Year	*Rising*	*Setting*
Spring (20 March)	90° (due E.)	270° (due W.)
Midsummer (20 June)	67° (23° N. of E.)	293° (23° N. of W.)
Autumn (20 Sept.)	90° (due E.)	270° (due W.)
Midwinter (20 Dec.)	113° (23° S. of E.)	247° (23° S. of W.)

APPENDIX 2

The sun is within a very few degrees of due S. at midday throughout the year. It is also higher in the sky at midday than at any other time, so that all shadows are then shortest. You can therefore take the shadow of a stick planted in the ground as pointing due N. when your watch tells midday, or when the shadow is shortest.

Appendix 3

THE CONSTELLATIONS

Many distances, of stars, clusters, nebulae, have been brought into line with recent measurements; but in view of the difficulty of measuring stellar distances these values have only a low level of reliability and are only of scientific value where they can be combined together for statistical analysis.

How to Find the Constellations

The four maps which accompany these notes show the appearance of the night sky at midnight on the first day of January, April, July and October, and on the other dates and times as stated. It will be noticed (*a*) that the constellations change with the seasons, as explained on pp. 23, 24, and (*b*) that the north and south points of each map are respectively at top and bottom, the east and west points on the left and right. To use one of the maps, therefore, it is most convenient to hold it above the head and to look up at it, facing south, when it will be correctly oriented.

Suppose the date is early January and the time midnight, and the reader wishes to identify the constellations that are visible. Using the appropriate map in the way described above, first find the Plough, which the map shows to be situated in the north-east sky. This star group is easily recognized and is always visible, being circumpolar; it is therefore a useful starting point for celestial explorations at any time of the year. The map shows that south of the Plough—i.e. in the south-east sky—lies Leo; this, with its first magnitude Regulus, will be spotted without difficulty. Carrying the eye westwards from Leo, the twin first magnitude stars Castor and Pollux of the constellation Gemini will be seen. Closely south-west of these is the striking constellation of Orion, with Sirius, the brightest star in the firmament, to the south-east (i.e. to the left of and lower in the sky than Orion). Thence Auriga and the W-

APPENDIX 3

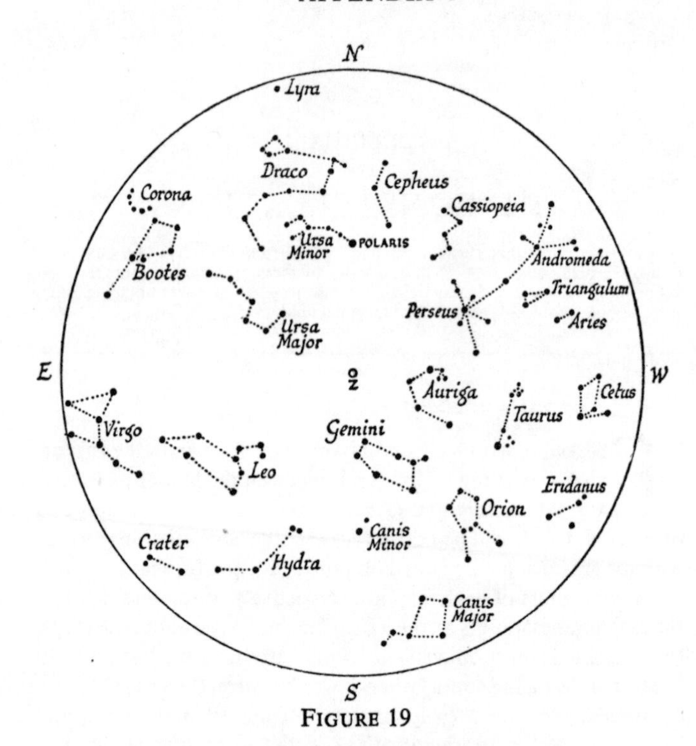

FIGURE 19

January 1st, 12 midnight; February 1st, 10 p.m., etc.

shaped Cassiopeia can be located. In this way more and more constellations are identified, the brighter and more conspicuous ones first, and the fainter ones being filled in afterwards.

Thus the entire dome of heaven can be covered, and even the complete beginner will find that after an hour's observation of this sort he will have familiarized himself with the appearance or 'shape' of the dozen or so bright constellations visible at the time. The procedure at any time of the year or any hour of the night is precisely the same.

Once this much has been achieved, the amateur astronomer will want to learn more about the constellations he has located. The

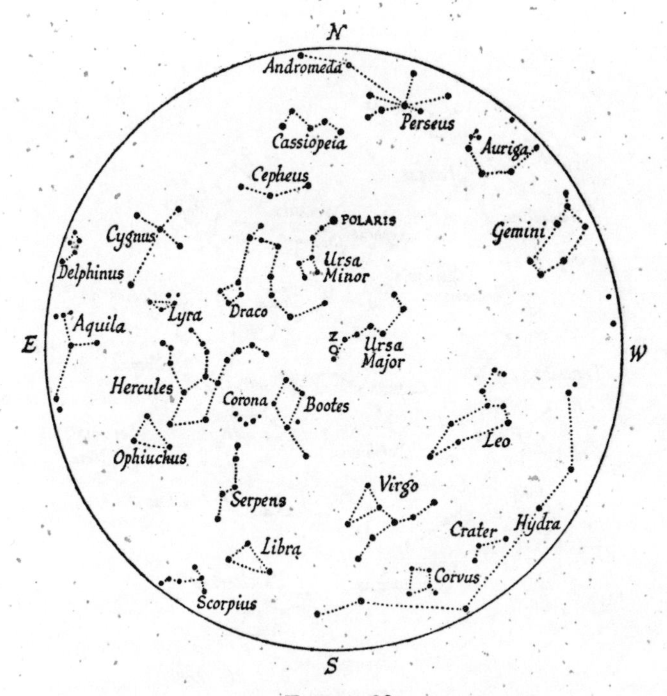

FIGURE 20

April 1st, 12 midnight; May 1st, 10 p.m., etc.

seasonal maps are therefore followed by detailed descriptions of each constellation in turn. These are arranged in order of Right Ascension—the astronomer's system of co-ordinates which corresponds on the star sphere to longitude on the earth's surface. In other words, the constellations are arranged in order from west to east, beginning with Pisces, the constellation in which the sun is to be found at the vernal equinox. For convenience of reference, they are preceded by an alphabetical Index.

Under each constellation will be found details of its position, a brief summary of its history and of the myths associated with it, a general description, particulars of such objects as are of special

APPENDIX 3

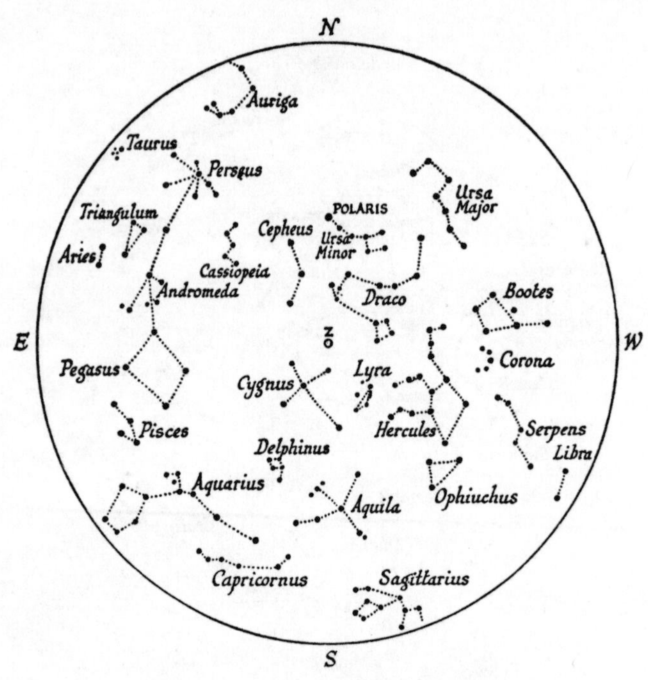

FIGURE 21

July 1st, 12 midnight; August 1st, 10 p.m., etc.

interest to observers with binoculars and the naked eye, and a large-scale map.

Stellar Nomenclature

The brighter stars of each constellation are designated by a Greek letter followed by the name of the constellation in the genitive case. Thus the brightest star in Orion is α Orionis ($= \alpha$ of Orion). When the Greek alphabet is exhausted in any constellation, the nomenclature of its stars is carried on with Arabic numerals. Variable stars are normally designated by capital letters—e.g. R Andromedae, SU Cassiopeiae, etc.

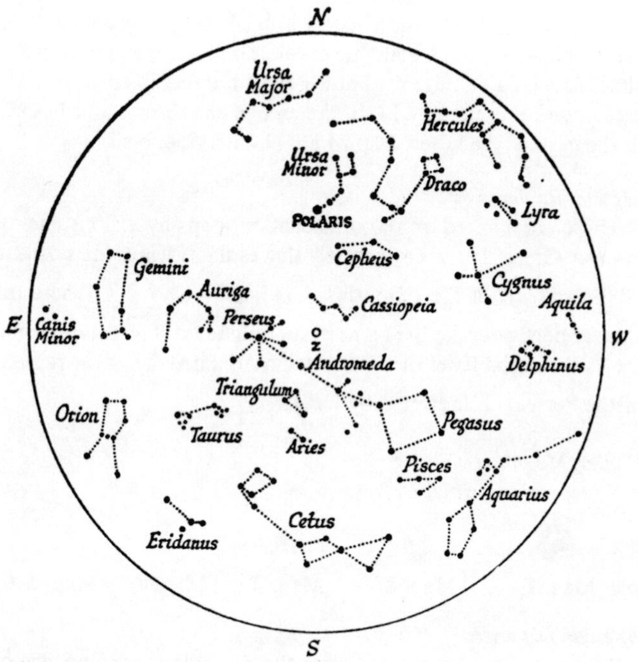

FIGURE 22

October 1st, 12 midnight; November 1st, 10 p.m., etc.

The Greek alphabet runs as follows:

alpha	α	iota	ι	rho	ρ
beta	β	kappa	κ	sigma	σ
gamma	γ	lambda	λ	tau	τ
delta	δ	mu	μ	upsilon	υ
epsilon	ϵ	nu	ν	phi	ϕ
zeta	ζ	xi	ξ	chi	χ
eta	η	omicron	o	psi	ψ
theta	θ	pi	π	omega	ω

Nomenclature of Nebulae and Clusters

The two standard catalogues of these objects are (*a*) the catalogue of 103 nebulae and clusters published by Messier in 1784, (*b*) the New General Catalogue of Sir John Herschel, published in 1888. A nebula or cluster is known by its number in either of these catalogues, preceded by M. in the case of the former and N.G.C. in the case of the latter. E.g. M.81, N.G.C.4254, etc.

Meteor Radiants

These are marked on the constellation maps by a circle enclosing the Greek letter designating the radiant. Thus the position of the radiant of the ζ Bootids is indicated by (ζ). Where the shower does not take the name of an individual star, the circle encloses the initial letter of the constellation name. Thus the radiant of the Perseids is indicated by (*P*).

Stellar Magnitudes

These are indicated as follows:

Angular Distances

To give the reader who is new to the constellations some idea of the apparent size of the star group he is looking for, the maps are provided with a scale. This scale, marked thus |▬▬|, represents a six-inch ruler held at arm's length against the sky. This is necessarily only a rough indication, since human arms are not all the same length, but it is accurate enough for some idea of the apparent size of the constellation to be gained from the map.

The six-inch ruler at arm's length subtends about 13°. The following angular measurements of the human hand when the arm is fully extended naturally vary to some extent from person to person, too, though they are useful rough guides:

Width of index finger-nail	1°
Distance across the knuckles of clenched fist	8°
Distance from 1st to 2nd knuckle	3°
Open span (thumb to little finger)	18°–20°

But the Plough provides a more accurate and varied set of sample measurements:

α to β (the Pointers)	5°
α to δ (the lip of the 'saucepan')	10°
Overall length of Plough (α to η)	25°
Polaris to nearer Pointer (α)	28°

The degree is divided into 60′ (minutes) and each minute into 60″ (seconds). Two stars which are about 30″ apart (1/1200 of the distance from α to δ Ursae Majoris) can be separated with average binoculars; this is the angle that a halfpenny subtends at about 200 yards.

Abbreviations

The following abbreviations are used in the lists of constellations and objects of interest:

L.Y.: light years.

Sp. bin.: spectroscopic binary.

L.P.V.: long period variable.

d.h.m.: days, hours, minutes.

⊙: the sun. Thus, 'Mass, 175 ⊙' means 'The mass of this star (or whatever the object may be) is 175 times that of the sun'.

Mag.: stellar magnitude.

Alphabetical List of Constellations

APPENDIX 3

PISCES: The Fishes

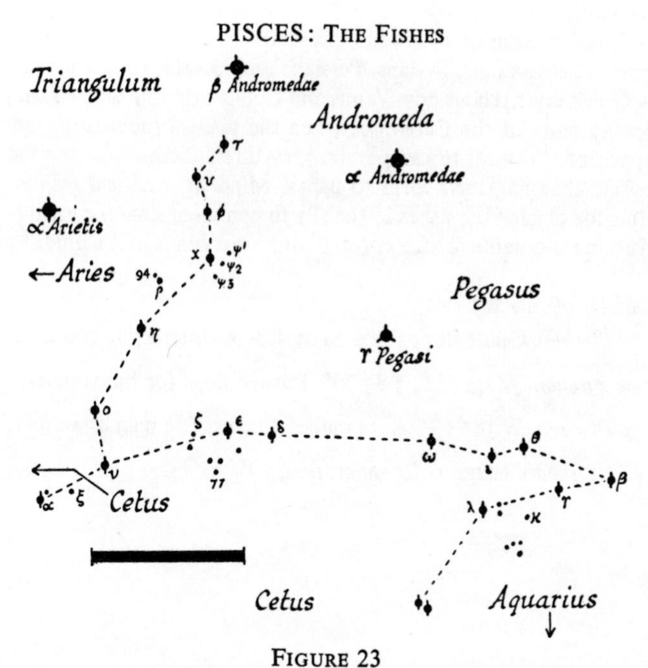

Figure 23

Position

Pisces is an autumn constellation, culminating at midnight[1] towards the end of September. It is sandwiched between Pegasus and Aries, with Cetus and Aquarius to the south, and Andromeda to the north.

Description

An inconspicuous constellation possessing no star brighter than mag. 4. It is the twelfth and last sign of the zodiac, the sun being in Pisces at the time of the vernal equinox; in other words, the celestial equator and the ecliptic intersect in this constellation.

[1] Midnight culmination: the approximate date on which the centre of the constellation lies on the meridian at midnight. This gives an indication of when each constellation should be looked for, i.e. whether it is a summer constellation, winter constellation, etc.

APPENDIX 3

Mythology and History

This constellation has been represented by two fishes from early times: Babylonians, Syrians, Persians and Greeks all saw it thus A Greek myth relates how Venus and Cupid were one day walking by the bank of the Euphrates when the giant Typhon suddenly appeared. In order to escape him they threw themselves into the water, changing their form to fishes. Minerva was held responsible for placing the fishes in the sky to commemorate the escape. Another old name for the constellation was Venus and Cupid.

Objects of Interest

κ *Piscium.* Guide star to an area of rich and interesting sweeping.

ψ *Piscium.* Mags. 5·6, 5·8; 30″. Rather close for binoculars.

ρ *Piscium.* With 94 Piscium makes a mags. 5, 6 naked-eye pair.

77 *Piscium.* Mags. 6, 7; separation, 33″.

ANDROMEDA

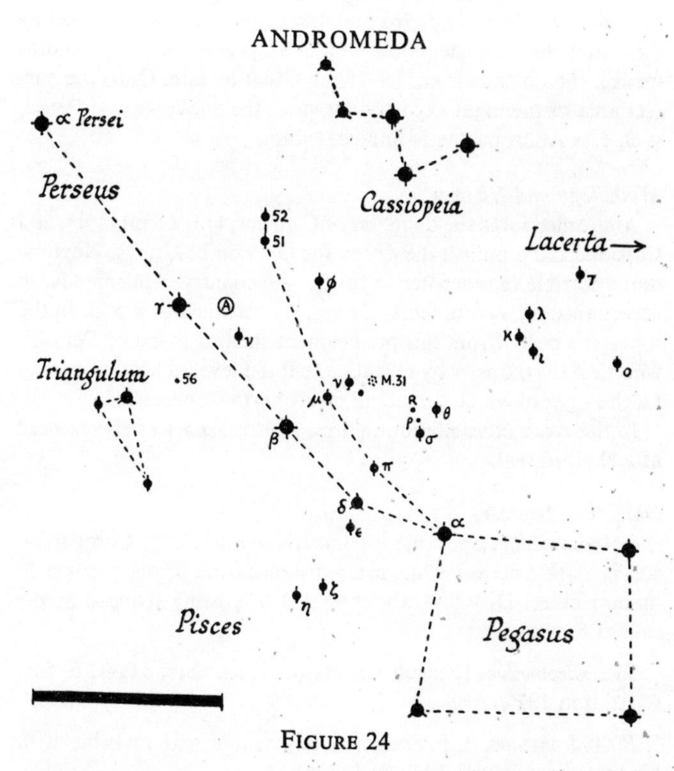

FIGURE 24

Position

Andromeda is an autumn constellation, culminating at midnight about October 10th; it is nevertheless observable, though nearer the horizon, for the greater part of the summer and early winter. It lies between Perseus and the Great Square of Pegasus, directly 'below' the W of Cassiopeia.

Description

The constellation contains no first magnitude star, but is easily identified when Perseus and Pegasus are known. It takes the form of a double row of 2nd, 3rd and 4th magnitude stars connecting the latter two constellations and converging upon α Andromedae, the north-east corner of the Great Square. Once the correct area of the night sky is under view, the line of stars α Persei, γ, β, δ, α Andromedae is unmistakable.

Mythology and History

Andromeda was the daughter of Cepheus, king of Ethiopia, and Cassiopeia. To punish the queen for her vain boastings, Neptune sent a terrible sea monster to ravage the country. Andromeda, in accordance with an oracular decree, was chained to a rock in the monster's path. From this predicament she was saved by Perseus, who slew the monster by revealing to it the severed head of Medusa, the sight of which turned all creatures to stone.

In the Arab constellation figures, Andromeda was represented as a chained seal.

Objects of Interest

α Andromedae. Sp. bin.; luminosity, about 175 ⊙. Companion star is 'dark'; period, 97d.; mean distance from primary, some 20 million miles. Distance (about 98 L.Y.) is being reduced at the rate of 8 m.p.s.

56 *Andromedae*. Difficult double for binoculars. Mags. 6, 5·8; separation 181″.

R Andromedae. L.P.V. Alternately visible and invisible with binoculars: mag. 5·6–14·9. Period, 410d.

M. 31, N.G.C. 244. The Great Andromeda Nebula. Largest and brightest of the spirals, and the only one visible to the naked eye (as a misty spot 1° west of the 4th mag. star ν Andromedae.) Discovered *c.* A.D. 950 by Sûfi, a Persian astronomer. Partially resolvable by the Mount Wilson 100-inch reflector into stellar points, some of which exhibit Cepheid variation. It is a galaxy of stars similar to our own stellar system, and is one of the few spirals that are approaching the sun, the velocity of approach being nearly 200 m.p.s. It appears, however, that this is due to the sun's galactic motion, and that the distance between our stellar system and the nebula is increasing, even though the distance between the sun and the nebula is at present decreasing. A number of novae have been discovered in M. 31; that of 1885 was mag. 7 at maximum. Distance, 1,800,000 L.Y.; diameter, 70,000 L.Y.; mass $3{\cdot}5 \times 10^{11}$ ⊙.

Andromedids. In the past this meteor swarm has provided brilliant showers (maximum, about November 27th), but now appears to be extinct. For connexion with Biela's comet, see p. 79.

APPENDIX 3

CASSIOPEIA

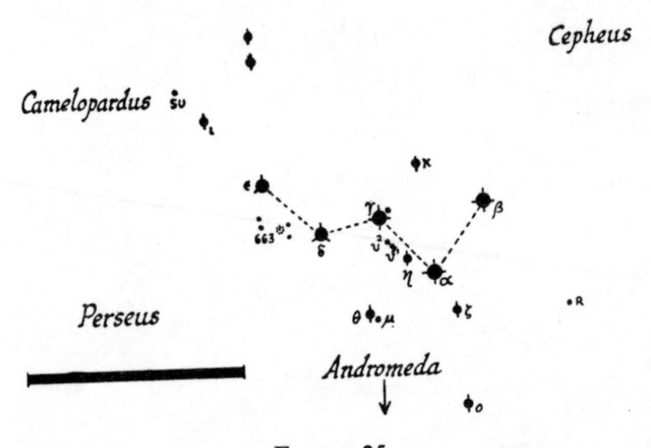

FIGURE 25

Position

Cassiopeia, being only 30° from Polaris, is in our latitudes a circumpolar constellation visible every night of the year. Its characteristic shape and position near the pole make it one of the easiest constellations to spot. It lies 'across' the north celestial pole from Ursa Major, and at about the same distance from Polaris as the Plough. γ Cassiopeiae, Polaris and ϵ Ursae Majoris lie in a straight line.

Description

The five main stars, magnitudes 2 and 3, form the outline of a rather distorted W. At times, of course, the W is lying on its side (when east or west of Polaris), and when above the Pole Star it is upside down—i.e. not a W but an M. The constellation itself extends for some distance round this asterism; about half of it lies within the confines of the Milky Way. There is interesting sweeping here, the star density being high, particularly around the middle of the W. The many beautiful clusters of Cassiopeia are unfortunately not accessible to binoculars.

Mythology and History

Cassiopeia was the wife of Cepheus, king of Ethiopia, and mother of Andromeda (q.v.). The following ancient designations of the constellation may be of interest, exhibiting as they do a much greater variety than is usual:

Egyptians: the Leg.
Greeks: the Laconian Key.
Romans: the Woman in the Chair.
Arabs: the Kneeling Camel, and the Lady in the Chair.
Esquimaux: the Stone Lamp.

The constellation, under one name or another, dates back to the fourth millennium B.C.

One of the most famous of all novae appeared in Cassiopeia in November 1572 and was observed by Tycho Brahe, whose recorded observations of the star are still in existence; it is commonly known as Tycho's Star. At maximum it was brighter than Venus, and was clearly visible during the daytime. By March 1574 it had faded below the limit of naked-eye visibility, having in the meantime changed to a deep red colour. Its approximate position was 2° from κ Cassiopeiae, along the produced line from η to κ. Two faint telescopic stars are now visible near this spot, but Tycho's observations were not accurate enough for us to be certain which of them, nearly 400 years ago, was the celebrated 'Blaze Star'.

Objects of Interest

α *Cassiopeiae*. Severe test object for binoculars. Mag. 2 reddish, mag. 9 blue; separation, 62″. Primary an irregular variable, period about 80d., mag. 2·2–3·1, discovered in 1831.

γ Cassiopeiae. Even more difficult. Mags. 2, 9; 432″. Distance about 125 L.Y.

ζ Cassiopeiae. Lies in a beautiful field of faint stars.

μ Cassiopeiae. A relatively near neighbour of the sun. Luminosity about 0·5 ☉. Large proper motion:[1] 100 m.p.s., equivalent to an angular displacement of more than 6′ per century. Radial velocity,[2] –60 m.p.s.

R Cassiopeiae. L.P.V. of striking deep red colour; mag. 5–12; period, 430d. Only visible intermittently with binoculars.

SU Cassiopeiae. Cepheid, entirely visible with binoculars. Mag. 5·9 6·3; period, 1d. 22h. 48m.

N.G.C. 663. Telescopically a fine galactic cluster; visible in binoculars.

[1] The proper motion of a star is its motion at right angles to the observer's line of sight. Directly measurable as an angular displacement on the star sphere, it can be expressed as a linear velocity as soon as the star's distance is known.

[2] The radial velocity of a star is its velocity of motion in the line of sight. A radial motion towards the earth is said to be negative, away from the earth positive. Thus 'radial velocity, –60 m.p.s.' means that the star is approaching us at a rate of 60 m.p.s.

CETUS: The Whale

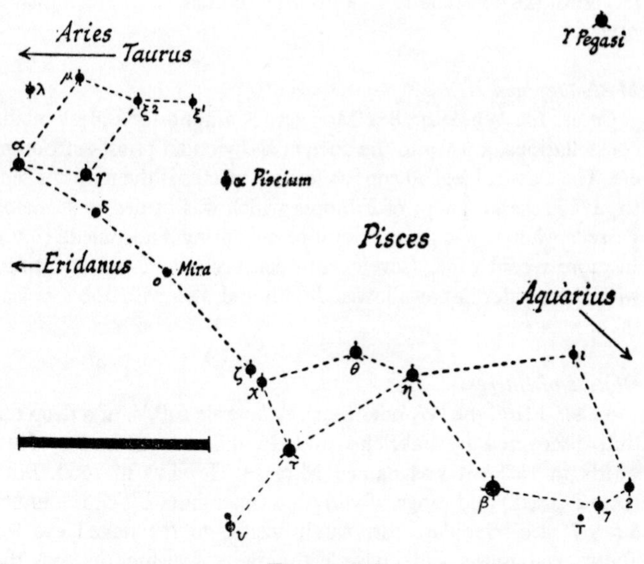

Figure 26

Position

The celestial equator runs through Cetus, a summer and autumn constellation which culminates at midnight about October 15th. It is a long constellation, bounded on the west by Pisces and Aquarius, on the north by Aries, and on the east by Taurus and Eridanus.

Description

Not one of the easiest constellations to identify. α Ceti, mag. 3, lies about 20° south-west of Aldebaran on the produced line from this star through λ Tauri. Thence the constellation straggles towards β Ceti, which lies near the westernmost end of Cetus, due south of α Cassiopeiae. It possesses one star of the 2nd magnitude

(β), and five of the 3rd; all the others are mag. 4 or fainter. With the exception of Mira, the wonderful variable (see below), this largest of the constellations holds little of interest for small instruments.

Mythology and History

Cetus, the Whale or Sea Monster, is among the oldest of the constellations, known to the ancient Babylonian priest-astronomers. The classical legend runs as follows: Cetus is the monster sent to ravage the kingdom of Ethiopia which was turned to stone by Perseus when it was on the point of devouring Andromeda (q.v.). In more recent times (seventeenth century) Cetus was identified with the whale that swallowed Jonah and also with Job's Leviathan.

Objects of Interest

o Ceti. Mira, the 'Wonder Star'. A notable L.P.V. of a deep red tint, discovered by Fabricius in 1596 and rediscovered by Holwarda in 1638; it was named Mira by Hevelius in 1660. Both period (331d.) and range of variation (maximum 1·7–5, minimum 8·5–9·7) are irregular. Alternately visible to the naked eye for about six months and invisible for about five months, but the whole of its fluctuation may be followed with binoculars. Duration of maximum is several weeks; waning occupies some eight months; the rise to maximum usually several weeks. Mira has a white dwarf companion, discovered in 1923 at a distance of 1″. Diameter of Mira A, about 400 ☉; of Mira B, 0·04 ☉. The primary is thus one of the largest stars yet investigated, and the comes one of the smallest.

The spectroscope has revealed the following facts:

(i) The star's temperature varies with the light variation: it is, paradoxically, hotter when decreasing in temperature than when brightening. It is deduced that its surface temperature at maximum is about 1,600° (c.f. the 6,000° of the sun).

(ii) Mira is approaching us at minimum and receding from us at maximum.

Distance about 125 L.Y.; mean radial velocity about +40 m.p.s.

χ *Ceti.* An easy test object for binoculars, the companion star being mag. 7·5.

T Ceti. Another variable that may be observed with binoculars; mag. 5·1–7·0. No definite period.

APPENDIX 3

TRIANGULUM: THE TRIANGLE

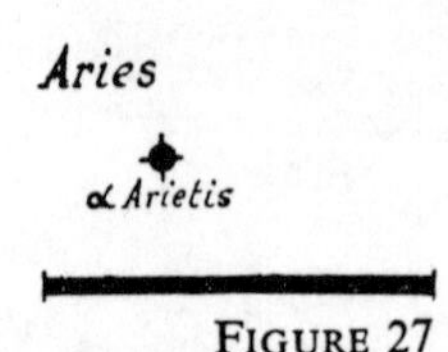

FIGURE 27

Position

An autumn constellation, culminating at midnight during the last week of October. It lies due south of Andromeda, and between that constellation and Aries. Its proximity with γ and β Andromedae (see map) makes it easy to find.

Description

The constellation consists of three stars (mags. 3, 4, 4·5) forming a small isosceles triangle lying on its side. The remaining stars are all faint.

Mythology and History

An ancient constellation, despite its smallness and lack of bright stars. The Greeks called it Deltoton, from its resemblance to their capital delta. The Hebrews named it after a triangular musical instrument. The stars α and β were called the Scale Beam by the Arabs. It was here that Piazzi discovered Ceres, the first asteroid, in 1800.

Objects of Interest

R Trianguli. L.P.V. visible with binoculars at and near maximum. Mag. 5·8–12. Period, 270d.

ARIES: The Ram

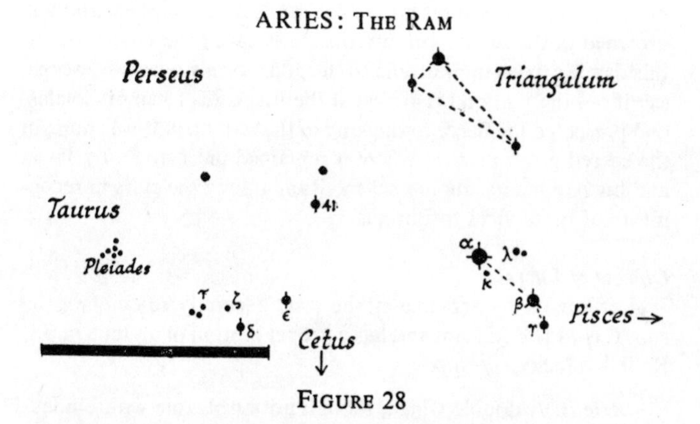

Figure 28

Position

An autumn constellation, being above the horizon at midnight from about June to February, culminating at the end of October. It lies south of Triangulum and Andromeda, and due west of the Pleiades in Taurus. To the west lies the constellation Pisces.

Description

Aries is the first sign of the zodiac. A small constellation possessing only five stars of magnitude 4 or brighter. α (mag. 2·2), β and γ (mags. 2·7 and 4·2) outline the naked-eye 'shape' of the constellation, a lopsided triangle.

Mythology and History

The constellation was named the Ram by Babylonians, Hebrews, Persians and Arabs. The Greek myth of the Golden Fleece contains a reference to the origin of the constellation. Phrixus and Helle, sons of the king of Thessaly, were ill-treated by their stepmother, and Mercury, taking pity on them, sent a ram with a golden fleece to help them to escape. The ram carried them away on its back, high above the earth, and all went well until they were crossing the strait separating Europe and Asia, when Helle momentarily released his grip of the ram's fleece, slipped, and was drowned in the sea far below—the sea is called the Hellespont to this day. Phrixus, in gratitude to the gods for his own safe escape, sacrificed the ram on his arrival at the Black Sea town of Colchis, and presented the fleece to the king of that country. It was hung in the sacred grove of Ares where it remained until stolen by Jason and his Argonauts. Jupiter set the Ram among the stars in recognition of its services to Phrixus.

Objects of Interest

α *Arietis*. A second magnitude star, generally resembling the sun. It is 74 L.Y. distant and has a proper motion of about 8 m.p.s. Radial velocity, –9 m.p.s.

γ *Arietis*. A double which, though not resolvable with binoculars (separation only 8″), is of historical interest as being one of the first telescopic doubles discovered—by Robert Hooke in 1664 while he was following a comet. Mags. 4·2, 4·4.

λ *Arietis*. A double with a separation of 38″; mag. 5 white, mag. 8 blue.

PERSEUS

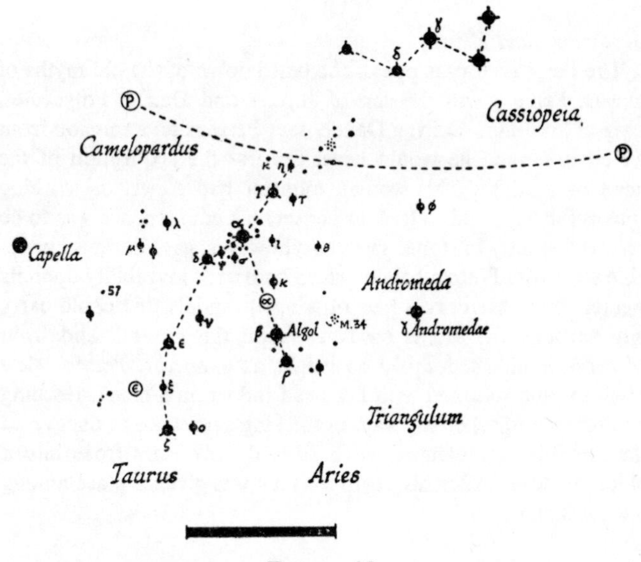

FIGURE 29

Position

Culminates at midnight about November 10th, when it is directly overhead, but it is high enough in the sky for midnight observation from July to March. The constellations that bound it are Andromeda and Triangulum to the west, Cassiopeia to the north, Auriga to the east, and Taurus to the south.

Description

Perseus is a compact constellation of 2nd and 3rd mag. stars and fainter, appearing to the naked eye as a curved line of stars running from Cassiopeia towards Capella. It lies for the most part within the confines of the Milky Way—extremely rich in this region, and well worth sweeping over—and possesses many objects of telescopic and binocular interest, including fine star fields (particularly near α), doubles, a superb cluster and a notable variable. The northern part of the constellation is circumpolar.

Mythology and History

The Perseus story is one of the best known of the old myths of Hellas. Perseus was the son of Jupiter and Danae. Polydectes, king of Scriphus, desiring Danae, sent Perseus on a mission from which it seemed he would never return—the acquisition of the head of Medusa. This woman monster had a nest of writhing snakes for hair, and to look at her terrible countenance was to be turned instantly to stone. The gods, however, gave Perseus invaluable aid: from Pluto a helmet which conferred invisibility upon its wearer, from Mercury a pair of winged sandals that could carry him through the air as easily as upon the ground, and from Minerva a polished shield as bright as a mirror. Perseus slew Medusa, and returned with her head hidden in a cloak, rescuing Andromeda (q.v.) on the way. In the king's presence he uncovered the horrible, snake-entangled head and Polydectes froze into a block of stone. After his death Perseus was given a place among the constellations.

Objects of Interest

α Persei. Situated in a beautiful field of faint stars.

57 Persei. Wide double, 114″; mags. 5 yellow, 6 bluish.

β Persei. Algol (Arabic: the Demon). Type star of the Algolid or dark-eclipsing variables. First detected by Montanari (1672); while Goodricke, an early and brilliant amateur observer, determined its period and suggested, correctly, that the variation might be due to the partial eclipses of a bright star by a dark or less luminous companion star revolving about it.

Some data of this interesting binary-variable are:

	Bright Primary	*Dark Companion*
Diameter:	3,107,000 miles	3,266,000 miles
Mass:	5.2☉	1.01☉
Separation:	6,830,000 miles	
Distance:	85 L.Y.	

The whole of the variation is visible to the naked eye. A steady maximum of mag. 2·3 is maintained for about 59h. with a fall of 0·05 mags. to a secondary minimum after 29½h.; it then fades to mag. 3·7 in a space of some 5h.; minimum is maintained for from 18–20m., when the rise to maximum is made in 5h. One complete cycle therefore occupies 2d. 20h. 49m., the period in which the two components revolve about their mutual centre of gravity.

ρ Persei. Irregular variable. Mag. 3·3–4·1.

M. 34, N.G.C. 1039. Loose galactic cluster, just visible to the naked eye on moonless nights, near the 'Double Cluster' (see below). Component stars average mag. 9. Distance, 1,700 L.Y.; diameter about 35′.

N.G.C. 869/884. The famous 'Double Cluster', clearly visible to the naked eye, and a superb spectacle in binoculars or a small telescope. The two clusters, each about 45′ in diameter, are of the loose galactic type and the finest specimens in the northern skies. They were well known to the ancients and are mentioned by Hipparchus and Ptolemy.

Perseids. One of the most prolific showers. August 4th–16th, maximum on or about August 12th, when as many as 50–70 meteors may be observed per hour. Position of the radiant moves appreciably throughout the duration of the shower (see map). Very swift and usually rather faint. Associated with Tuttle's comet 1862 III. Earliest records of this shower date back to A.D. 811. The Perseids used to be known as the Tears of St. Lawrence, since they are to be seen around the date dedicated to that saint.

α–β Perseids. A less conspicuous shower of very swift meteors. July 25th–August 4th.

ε Perseids. Also less important than the July–August Perseids. Swift moving. September 7th–15th.

APPENDIX 3

ERIDANUS: THE RIVER

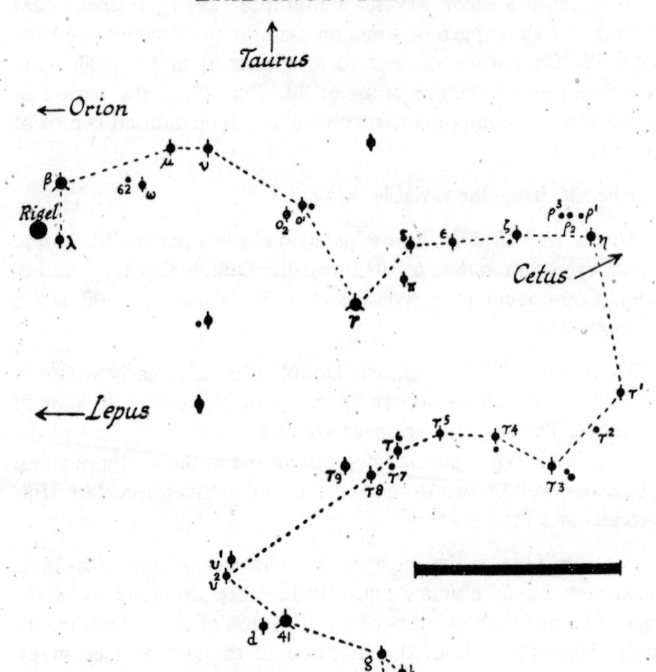

FIGURE 30

Position

A winter constellation, culminating at midnight about November 10th. It lies west and south-west of Orion, south of Taurus, and east of Cetus. Its northern boundary is the celestial equator due east of δ Orionis, whence it straggles southward until it passes out of sight below the southern horizon.

Description

Eridanus is a long, wandering constellation of which only the northern part is visible in these latitudes. Many of its brighter stars are never visible in the British Isles, including the 1st mag. α, which is only 30° from the south celestial pole.

History and Mythology

An old constellation usually named by the ancients after their local river. Thus to the Egyptians it was the Nile, the Euphrates to the Babylonians, and the Jordan to the Jews. The Arabs knew it as the River simply. The name Eridanus is of Greek origin.

Objects of Interest

β Eridani. Marks an area worth sweeping over, on the Orion border.

o Eridani. Naked eye pair, mags. 4, 4·5. o^2 has a faint companion, very near or below the reach of good binoculars: mag. 9·4, separation 85″. This binary, 15 L.Y. distant, was discovered by Herschel in 1783. A third companion is invisible in small instruments.

62 Eridani. Faint double: mags. 6, 8; separation, 63″.

APPENDIX 3

TAURUS: THE BULL

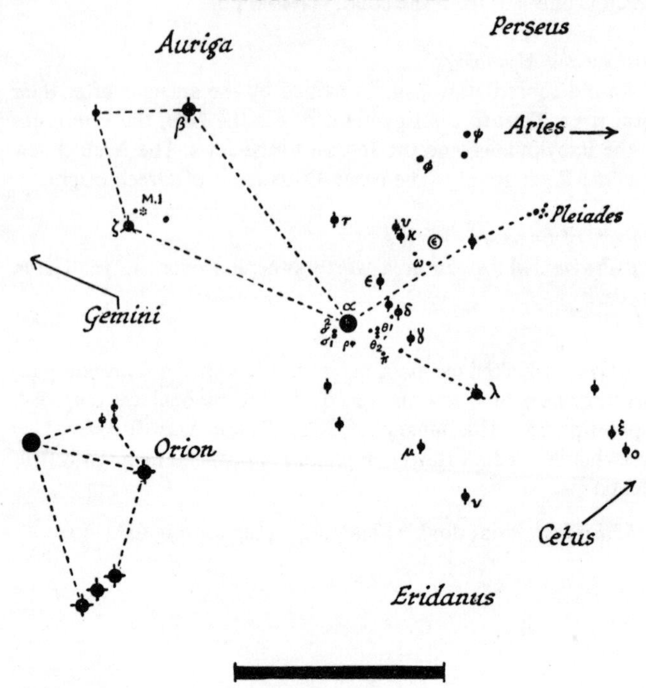

FIGURE 31

Position

A fine winter constellation, culminating at midnight at the beginning of December. It is bounded by Orion and Gemini on the east, Auriga and Perseus on the north, Aries and Cetus on the west, and Eridanus on the south. It is easily spotted by the naked eye from its position closely north-west of Orion, and from the two star clusterings, the Hyades and the Pleiades, which it contains.

Description

α Tauri is a reddish, first magnitude star situated on the eastern edge of the coarse clustering known as the Hyades. The well-known Pleiades lie some 15° to the north-west. Taurus is the second sign of the zodiac.

Mythology and History

A very ancient constellation, probably first delineated by the Chaldeans or the Egyptians between 4,000 and 1,700 B.C. The Egyptians associated the constellation with Osiris, who in addition to being the bull-god was the special god of the Nile. The constellation was also known as the Bull by the Hebrews. A Greek legend relates how Jupiter, burning with desire for Europa, the beautiful daughter of the king of Phoenicia, assumed the form of a snow-white bull and mingled with her herd. Europa, struck by the beauty of the animal, stroked and patted it and finally mounted its back. The bull immediately carried her off across the seas to Crete, where it reassumed its divine form before Europa's astonished eyes.

Mythology of the Hyades

Two legends refer to this beautiful clustering of stars which lies closely west of Aldebaran, the eye of the Bull. The first identifies the Hyades with the nurses of the infant Bacchus, who on their death were rewarded for their services by being translated to the stars. The second sees them as the five daughters of Atlas who pined away and died of sorrow over the death of their beloved brother Hyas, who had been gored by a wild boar. The ancients always associated the Hyades with rain, possibly in memory of the tears of Atlas' daughters, but more probably because the first sight of these stars in the autumn heralds the rainy season.

Mythology of the Pleiades

This cluster has excited wonder and interest from the earliest times and was an object of veneration by the Chinese (in whose writings mention of it occurs as early as 2357 B.C.), Egyptians, Japanese, Hindus and Aztecs. Mention also occurs in Sappho (600 B.C.), Euripides (500 B.C.) and the Bible. In classical mythology the Pleiades are the half-sisters of the Hyades, daughters of Atlas and Pleione, placed among the stars in recognition of their filial affection and sorrow for their father, who had to bear the world upon his shoulders. Their names are Alcyone, Merope, Maia, Electra, Sterope and Celaeno.

Objects of Interest

α Tauri. Aldebaran. Mag. 1 reddish (compare it with the remarkably white β Tauri). A mag. 11 comes is not visible with binoculars. Diameter, about 35☉; brightness, 90☉; distance, 68 L.Y.; radial velocity, +34 m.p.s.

θ Tauri. Easy naked-eye double, worth viewing with binoculars; mags. 4, 4; separation, 5′ 30″. θ^1 (the southernmost) is greenish white, θ^2 yellowish.

σ^1, σ^2 Tauri. Another wide double. Close to Aldebaran. Mags. 5, 5; separation, 7′.

φ Tauri. Difficult, owing to faintness of comes. Mag. 5 reddish, 8 blue; separation, 50″.

τ Tauri. Another pair with widely differing magnitudes: mags. 4, 7, white and blue; separation, 63″.

λ Tauri. Algolid (dark-eclipsing) variable. Mag. 3·4—4·3. Period, 3d. 22h. 52m.

Pleiades. The finest of the very open clusters; somewhat reminiscent of a miniature Plough. Six or seven stars will probably be seen with the naked eye, though as many as eleven, twelve, and even fourteen have been claimed without optical aid. Of the thousands of faint stars shown by photography, probably only the brighter are associated with the cluster. Distance about 400 L.Y.; some 12 L.Y. from end to end. The brighter components have a common proper motion of about 6″ per century towards Betelgeuse. Photography has revealed that the whole cluster is involved in diffuse nebulosity. The faint nebula entangling Merope is also invisible in small instruments.

Hyades. More open than the Pleiades; best seen with binoculars. A large number of the Hyades stars have an identical proper motion with that of the Pleiades—i.e. towards Betelgeuse in Orion. In addition, the majority have positive radial velocities. This general motion is not shared by Aldebaran, which is unconnected with the cluster, being very much nearer the sun. Distance about 125 L.Y. (Aldebaran, 68 L.Y.).

M. 1, N.G.C. 1952. The 'Crab Nebula', visible in binoculars as a minute point of light closely W. of the isosceles triangle of faint stars which is itself closely W. of ζ. Discovered in 1731, and rediscovered in 1758 by Messier, who made it the first entry in his famous catalogue of 103 nebulae and clusters.

Taurids. Radiant near ϵ provides occasional meteors from October 26th to November 16th approximately; hourly rate of 5 or 6 during maximum, November 4th–9th. Typically slow moving.

APPENDIX 3

ORION

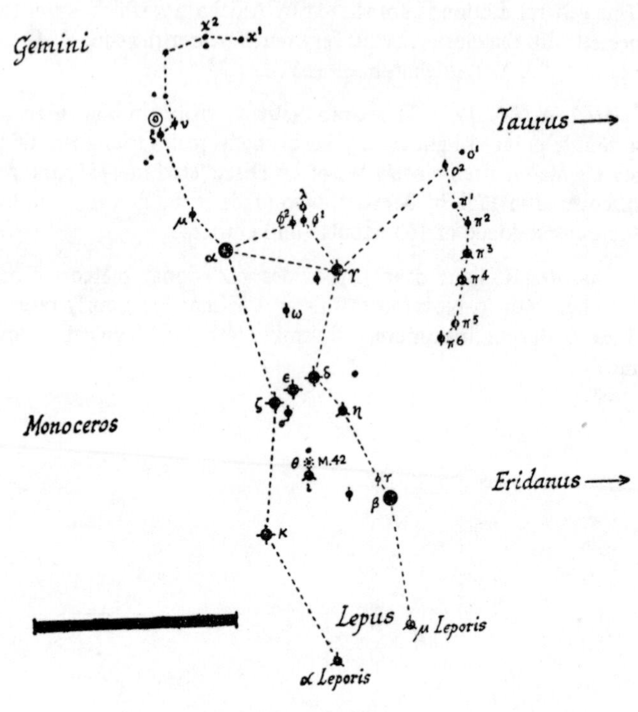

FIGURE 32

Position

To describe the position of this constellation with reference to the surrounding constellations is rather like describing the location of London by saying that it is south of Pinner, north of Kingston, east of Osterley, etc. For Orion is far and away the most prominent star group of the winter skies, outshining all its neighbours in impressiveness. Actually it lies east of Taurus and Eridanus, north of Lepus, west of Monoceros and Gemini. Conspicuous to the north-east and south-east are Procyon (α Canis Minoris) and

Sirius (α Canis Majoris) respectively. δ Orionis lies almost exactly on the celestial equator. The constellation culminates at midnight in mid-December.

Description

Orion is the finest of the winter constellations, perhaps the finest visible in our northern skies. He is the Hunter, and it is easy to see his head and broad shoulders, his belt with a dagger hanging from it, his knees and (in the stars α and μ Leporis) his feet. To his west a curved line of six stars (π^1–π^6, mags. 3, 4 and 5) represents the lion's skin he carries over his left arm, whilst with his right hand he brandishes an upraised club, terminating in the 5th mag. χ^1 and χ^2. On clear winter nights, Orion striding across the frosty sky, resplendent with Betelgeuse and Rigel and followed by his two dogs (Sirius and Procyon), is a fine spectacle. The constellation is worth sweeping over carefully with binoculars.

Mythology and History

This star group is of the greatest antiquity, and was recognized by the astronomers of the Euphrates valley at least as early as 2000 B.C. and possibly even 2,000 years before that. The Chaldeans called it Tammuz, after the month of that name (June) during which, at the former date, the Belt stars first became visible above the eastern horizon before sunrise. Both Syrians and Arabs called the constellation the Giant, while the Egyptians saw it as Horus, the young sun god. In Greek mythology Orion was the son of Neptune and the Amazon Queen Euryale. He inherited his mother's hunting and fighting skill to the extent of earning for himself the reputation of being the greatest hunter in the world. To punish the pride that sprang from this supremacy over all other men the gods sent a scorpion which bit him in the foot and caused his death. Diana placed him in the skies directly opposite the Scorpion, where he could never again suffer any harm from it.

Objects of Interest

α *Orionis*. Betelgeuse. Irregular variable, mag. 0·5–1·4; its variability was first noticed by the younger Herschel a hundred years ago. One of the reddest of the brighter stars, and the first star whose diameter was directly measured by the Mount Wilson interferometer (1920). Betelgeuse is actually pulsating (no doubt the cause of the light variation), its mean diameter being 300 million miles, i.e. greater than the orbit of Mars, or 350$\odot$. Volume, 27,000,000$\odot$; mass, 35$\odot$; density, 0·001 times that of air; distance, 587 L.Y.

β Orionis. Rigel. One of the intrinsically brightest stars yet investigated. Luminosity, 14,000$\odot$; mag. 0·3; distance, 870 L.Y.; diameter, 35$\odot$; radial velocity, +14 m.p.s. One or two faint companions are visible with binoculars.

δ Orionis. Mags. 2, 6·8, white; separation, 52″. The primary is a slightly variable sp. bin. Probable distance, 1,470 L.Y.; luminosity, 3,000$\odot$; radial velocity, +12 m.p.s.

λ Orionis. Brightest star in the 'head' of Orion; repays careful binocular scrutiny.

π^5 Orionis. Indicates position of a mag. 6 red star, 15′ N.W.

θ Orionis. Central star of Orion's dagger. Fine quadruple star, commonly known as the Trapezium, unfortunately not visible as such in binoculars. The components are of mags. 6, 7, 7·5 and 8. θ lies in the centre of the Great Orion Nebula (see below).

M. 42, N.G.C. 1976. The Great Orion Nebula. In large instruments it is an incomparably grand object—a gigantic, convoluted cloud of incandescent gas. Long exposure photography has extended its ramifications over a great part of the constellation. It is plainly visible to the naked eye as a misty spot, and even in binoculars is an unusual object, a vague mist of pale green light. The bright central region near θ is some 8 L.Y. across, and its mass may be 10,000$\odot$, though this is necessarily a very rough estimate. Distance, about 1,500 L.Y. M. 42 was first noticed, so far as is known, by Cysatus in 1618, though he was unaware of its true nature.

Orionids. October 15th–25th; maximum, October 21st, when some twenty swift-moving meteors per hour may be expected. The radiant is located close to ξ Orionis.

LEPUS: The Hare

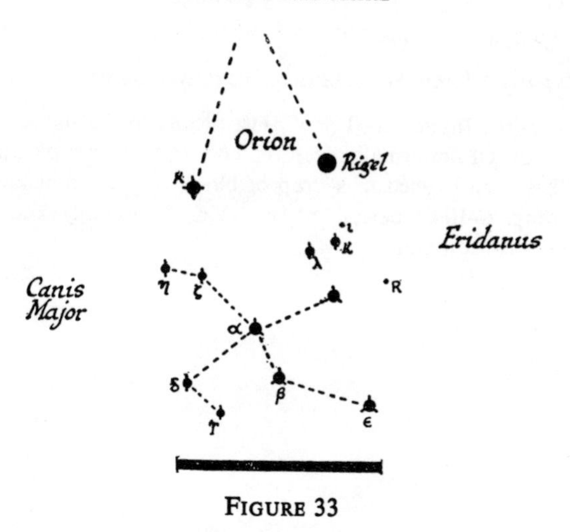

Figure 33

Position

A winter constellation, culminating at midnight in mid-December. It lies due south of Orion, west of Canis Major and east of Eridanus.

Description

A small constellation, easily found from its proximity with Orion. It is moderately conspicuous, having four mag. 3 stars, a further six being brighter than mag. 5.

Mythology and History

An ancient constellation, though it has been known by a variety of names. To the Egyptians it was the Boat of Osiris. The Arabs thought of it as Orion's Chair. The name Lepus is due to the Romans, and it was said that since the hare was the animal that Orion particularly liked hunting, one was placed near him in the sky.

APPENDIX 3

Objects of Interest

γ Leporis. Mags. 3·8 yellow, 6·4 reddish; separation, 95″. A third companion is invisible with binoculars.

ι Leporis. Closely W. lies a mag. 5·5 deep red star.

R Leporis. Interesting L.P.V., just visible to the naked eye at maximum. Of abnormally deep red colour, its telescopic appearance has been likened to a drop of blood suspended against the sky. Mag. 6–10·4; period, about 420d. Commonly known as Hind's 'Crimson Star'.

AURIGA: THE CHARIOTEER

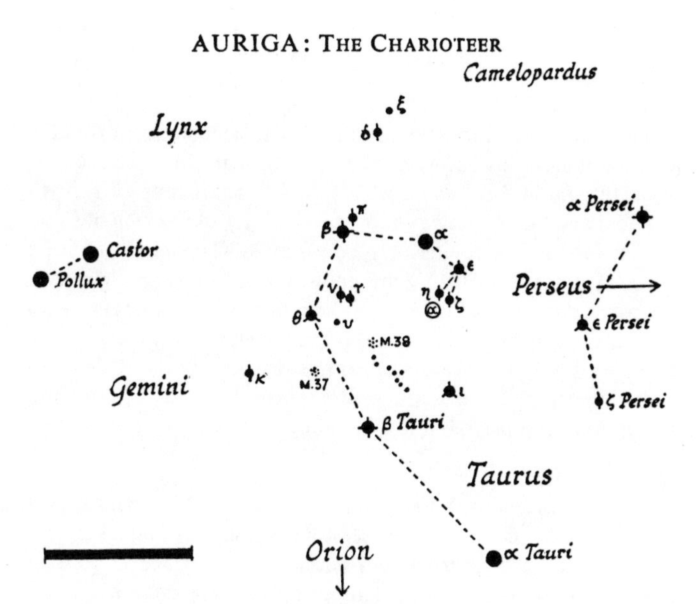

FIGURE 34

Position

A winter and spring constellation, culminating at midnight a few days before Christmas. It lies north of Orion, north-east of Taurus and the Pleiades, east of Perseus, and west of Gemini.

Description

Capella (see below), a beautiful yellow star of the 1st magnitude, is the lucida of the constellation. Auriga is conspicuous and easily found: the curved line of stars, β Tauri (situated on the boundary of the two constellations), θ, β and α Aurigae, ending in the triangle of fainter stars known as the Kids, is unmistakable when once found, and may be thought of as resembling a drooping, heavy-headed flower. The Milky Way passes through the western reaches of the constellation, Capella lying on its edge. ι Aurigae lies on the further edge of the Milky Way and marks an area of fine sweeping; the whole of the pentagon bounded by the above-mentioned stars repays sweeping over, for it contains some fine clusters (see below) and interesting groupings of faint stars.

Mythology and History

The constellation is referred to in two ancient stories. From the earliest times it has been pictured as a man carrying a goat on his shoulder and two kids in his left hand. According to the first myth the goat and kids commemorate the daughters of the king of Crete who fed the infant Jupiter with goat's milk. The second tradition is that Auriga represents Erichthonius, son of Vulcan and Minerva. He was a cripple, and in order to help himself to get about he invented the four-horse chariot. To immortalize this invention the first Charioteer was translated to the skies. Mention of the constellation occurs in the writings of Aratus and Eudoxus (third and fourth centuries B.C.), but it is certainly older than this.

Objects of Interest

α Aurigae. Capella, a fine yellow star, mag. 0·2, visible for part of every night of the year. Ptolemy (A.D. 150), Al Fagani (tenth-century Arab astronomer), and Riccioli all described Capella as red; it is possible that it has changed colour in more recent times. Sp. bin.; period, 104d.; the first binary of this type to be resolved by the Mt. Wilson interferometer. Distance, 46 L.Y.; radial velocity, +19 m.p.s. Mass 5☉, diameter 9☉, luminosity 80☉ (approximate).

β Aurigae. Interesting sp. bin. Period, 3d. 23h. 2m.; diameters both 3☉; separation of centres of the two components, 7,650,000 miles; mass of A, 2·2☉; mass of B, 2·1☉; distance, 85 L.Y.

M. 37, N.G.C. 2099. Fine open cluster, visible with good binoculars.

M. 38, N.G.C. 1912. A loose cluster, worth searching for with good glasses. A telescope shows it to be roughly cruciform. Even if invisible, the sweeping hereabouts is very fine.

α Aurigids. February 5th–10th; very slow moving. A second shower of the same name from a nearby radiant occurs between August 12th and October 2nd; swift moving. Both are minor showers.

CANIS MAJOR: THE LARGER DOG

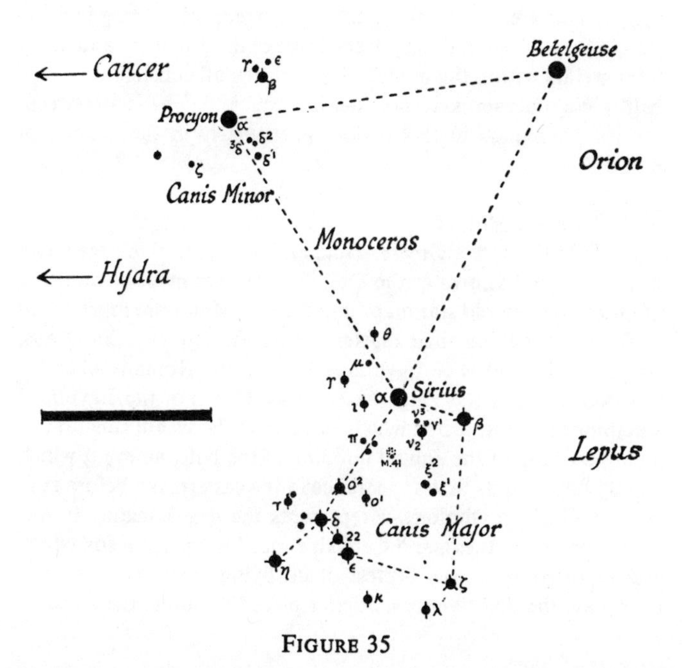

FIGURE 35

Position

One of the fine winter constellations, culminating at midnight about New Year's Day. Canis Major lies roughly 10° to 30° south of the celestial equator and therefore requires to be observed when not too far from the meridian. It lies south-east of Orion, where the brilliant Sirius makes it quite unmistakable. It is bounded on the north by Monoceros, on the east and south by the southern constellation Argo, and on the west by Lepus. Sirius itself lies almost due south of the centre of the line joining α Canis Minoris (Procyon) and α Orionis (Betelgeuse).

Description

A small but notable constellation. Sirius (see below) is the brightest star in our skies and a brilliant object on frosty winter nights. This area of the sky is thickly starred, there being in addition to Sirius, four 2nd mag. stars, three of the 3rd mag., and many fainter stars within the boundaries of the small constellation. The Milky Way cuts across its north-east corner and there is fine sweeping for binoculars in this region, particularly in the vicinity of η and δ.

Mythology and History

Canis Major is the Greater Dog, and Sirius has from the dawn of history been known as the Dog Star. Homer mentions the dog of Orion, and on old star maps it and Canis Minor are represented as following behind their master, Orion the Hunter. Chaldeans, Assyrians, Phoenicians, Egyptians, Greeks and Romans all called this constellation the Dog. As early as 3285 B.C. the Egyptians worshipped Sirius, a star which performed the useful function of giving warning of the annual flooding of the Nile, an event which closely followed its first appearance in the eastern sky before sunrise. In Greek mythology it represents the dog Laelaps. It was noted for its swiftness, and Cephalus raced it against a fox which was reported to be the fleetest of all living creatures. To commemorate the dog's victory, Jupiter placed it among the stars.

Objects of Interest

α *Canis Majoris*. Sirius, mag. –1·6; distance, 8·6 L.Y.; luminosity, 27☉; diameter, 2☉; temperature, 10,000° (1·6 ☉); radial velocity, –5 m.p.s.; proper motion, 11 m.p.s. Visible in daylight with a ½-inch glass, and casts a perceptible shadow on moonless nights. One of the most interesting binaries in the sky. Its white dwarf companion was discovered in 1862 by Alvan Clark while testing a new 18½-inch objective; before this time irregularities in the proper motion of Sirius had led Bessel and Auwers to predict such a discovery. Following are some of the data of Sirius B: mag. 8·4; separation from Sirius A, 2″–11″; period, 48 years; brightness, 0·0001 times Sirius A; size, about 3 times earth; mass, 250,000 times earth, 0·4 times Sirius A; density, 36,000⊙, i.e. 5,000 times as dense as lead or 60,000 times that of water. In other words, 1 cubic inch of the matter of Sirius B weighs a ton, and although its diameter is only about 26,000 miles it contains approximately as much matter as the sun (diameter =864,000 miles). It is thought possible that the colour of Sirius may have changed within historical times, for Seneca, Ptolemy and other ancient writers and observers described it as red.

β Canis Majoris. A mag. 2 star, called by the Arabs 'the Herald of Sirius'. Its luminosity is about 2,000⊙, distance 650 L.Y., radial velocity +20 m.p.s.

ν Canis Majoris. Triple; a beautiful sight in binoculars.

22 Canis Majoris. Compare the red tint of this mag. 3·5 star with the nearby ε (white).

δ Canis Majoris. Note the train of faint stars curling round this star towards ω^1 and ω^2. Fine sweeping hereabouts.

M. 41, N.G.C. 2287. A very fine cluster, visible to the naked eye 4° south of Sirius. Even binoculars show it as a wonderful object. A fine red star at the centre of the cluster is unfortunately invisible with apertures less than about 5 in.

APPENDIX 3

GEMINI: The Twins

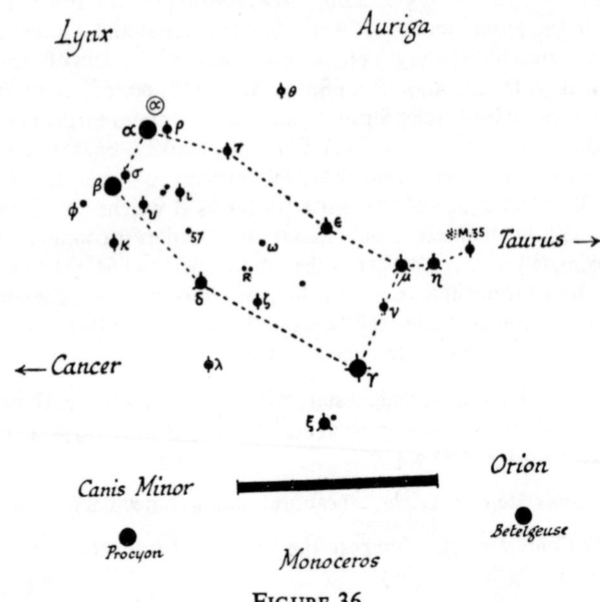

FIGURE 36

Position

A winter constellation, culminating at midnight early in January. It lies north-east of Orion, east of Taurus, and west of Leo; south of it lies Canis Minor.

Description

Gemini is a conspicuous group of 1st, 2nd and 3rd mag. stars forming a rough quadrilateral figure. Its south-west corner lies in the Milky Way and there is fine sweeping within its boundaries, notably in the region of η and ϵ. Gemini is the third sign of the zodiac.

Mythology and History

This constellation has been known as the Twins (twin men,

plants, deities or animals) from the earliest times and by widely different cultures and races. In classical mythology the twins are Castor and Pollux, whose names have been given to the stars α and β. They were the sons of Leda, wife of the king of Sparta, by Jupiter (see under Cygnus), and sailed with Jason on the Argo in search of the Golden Fleece (see under Aries). They were invincible fighters, and inseparable companions. To commemorate their prowess and their love for one another Jupiter, their father, transported them to the skies after their death. On account of the help they gave their fellow Argonauts when a storm threatened to sink their ship, the constellation was in early times thought to be a favourable sign for sailors. It will be remembered that the ship in which St. Paul sailed from Malta was named the *Castor and Pollux.*

It was in this constellation that both Uranus and Pluto were discovered: the former near η by Herschel in 1781, and the latter near δ by Tombaugh in 1930.

Objects of Interest

α Geminorum. Castor. One of the loveliest doubles in the heavens, but too close (3″·9, closing) for binoculars. A true binary, period about 350 years. Both components are sp. bins., periods 2·9d. and 9·2d. A faint red companion 73″ distant, also a member of the system, is itself a close eclipsing binary, period 0·8d. Castor is thus a sextuple star. Distance, 45 L.Y. Other data:

	Castor A	*Castor B*	*Castor C*
Brightness:	23 ☉	10 ☉	0·04 ☉
Mass:	5·5 ☉		*c.*3 ☉

Note contrasted colours of Castor (white) and Pollux (deep yellow). Both stars have several optical companions.

It was the striking appearance of Castor that led Herschel to conceive of true binaries; previously all close pairs of stars had been thought to be no more than optical doubles.

ζ Geminorum. Wide double (94″), mag. 4 yellow, mag. 7 blue. Primary is a Cepheid; period, 10d. 3h. 43m.; mag. 3·7–4·5.

η *Geminorum.* L.V.P.; period, 230d. Mag. 3·2–4·2. Fine sweeping here and towards ϵ.

R Geminorum. L.P.V. (370d.); mag. 6·0–13·8. Only visible with binoculars near maximum.

M. 35, N.G.C. 2168. Fine loose cluster, just visible to the naked eye, 2° north-west of η. 500–600 stars of mags. 9–16. Fine sweeping along the edge of the Galaxy in this region; note the two lines of faint stars running towards the cluster from μ and η.

Geminids. Important shower which reaches maximum about December 13th; occasional meteors visible throughout the month. As many as 60 bright, swift moving meteors per hour may be seen at maximum.

MONOCEROS: The Unicorn

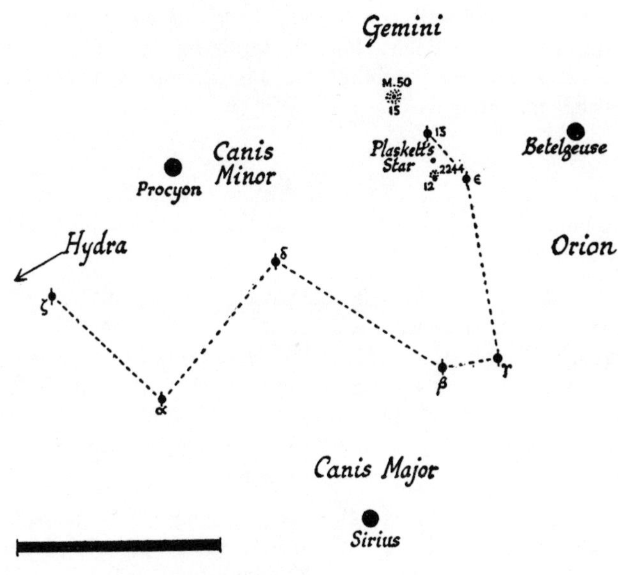

Figure 37

Position

An autumn to spring constellation, culminating at midnight on about January 5th. It lies to the north of Sirius and Canis Major, and below Orion and Canis Minor; on the north it is bounded by Gemini. It is bisected by the celestial equator.

Description

This inconspicuous constellation lies almost entirely within the region of the Milky Way and although of small interest for the naked-eye observer abounds in star fields which are marvellous sights even with binoculars, as well as in many objects of telescopic interest. It possesses no star brighter than the 4th magnitude.

APPENDIX 3

Mythology and History

A comparatively recent constellation, certainly added later than the time of Ptolemy (fl. *c.* A.D. 150), though its precise date and creator are unknown. Bartschius, the son-in-law of Kepler, is one claimant to the title. Monoceros is a mythological or heraldic creature resembling the unicorn.

Objects of Interest

ε Monocerotis. A 4th mag. double, too close for separation with binoculars. Marks a fine area for low-power sweeping, such as is provided by binoculars.

1309 Monocerotis. Plaskett's Star. A close binary (mag. 6) with some of the most unusual features of any star yet investigated. Period, 14½d.; separation, 56,000,000 miles (5/8 earth-sun); orbital velocity, 155 m.p.s.; mass of primary 76⊙, of comes 63⊙; luminosity, about 30,000⊙. These two stars are the most massive known; stellar masses approaching even 50⊙ are exceedingly rare.

M. 50, N.G.C. 2323. Nebulous cluster, some 20′ in diameter, containing the variable 15 Monocerotis (mag. 4·9–5·4; period, 3½d.). Estimated distance of M. 50, more than 1,000 L.Y.; diameter about 8 L.Y. Fine sweeping hereabouts.

N.G.C. 2244. Fine cluster (estimated distance, 350 L.Y.), just visible to the naked eye on dark nights, and a glorious sight in binoculars. Mags. 6–14, majority 6–8, including the mag. 6 star 12 Monocerotis. This, however, is probably very much nearer than the cluster and unconnected with it.

CANIS MINOR: The Smaller Dog

For constellation map, see Fig. 35

Position

A winter constellation, midnight culmination occurring in the middle of January. It lies on the celestial equator east of Orion, south of Gemini and south-west of Leo.

Description

A small constellation, distinguished only on account of the 1st mag. star Procyon, α Canis Minoris. β is mag. 3, but it has no other stars brighter than the 5th magnitude.

Mythology and History

The Romans called this constellation the Puppy, the Arabs called it the Lesser Dog, while the Greeks and Egyptians likewise saw the figure of a dog in this region of the winter sky. There are various myths and traditions relating to the identity of the Smaller Dog. One tells that it was one of the hounds of Actaeon which, after Diana had turned him into a stag for surprising her while she was bathing in a woodland pool, turned upon and tore to pieces their erstwhile master. It may originally have been Anubis, the dog-headed god of the Egyptians, or one of the hounds of Diana. The most reasonable supposition, however, is that it and Canis Major are the hounds of Orion the Hunter, whom they eternally follow across the skies.

Objects of Interest

α Canis Minoris. Procyon, mag. 0·4; a lovely, deep yellow star. Luminosity, 7☉; radial velocity, –2 m.p.s.; proper motion, 2′ per century. A binary system in some ways resembling that of Sirius. Procyon B, discovered by Schaeberle in 1896, is only visible in moderately large telescopes; mag. 14, separation 44″.

	A	*B*
Brightness:	5☉	0·0003☉
Mass:	0·25☉	0·3–0·25☉
Period:		39 yrs.

β Canis Minoris. Mag. 3. Binoculars show two faint companions, one noticeably reddish.

APPENDIX 3

LYNX

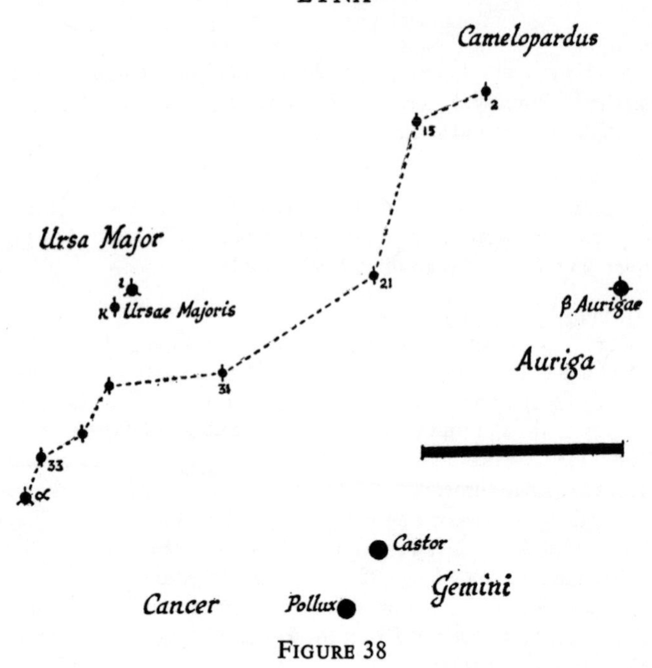

FIGURE 38

Position and Description

A large and inconspicuous constellation containing no object of particular interest for the observer with a small glass, although it is rich in doubles suitable for telescopic viewing. It is partially circumpolar, and culminates at midnight about the middle of January. Its boundaries are:

On the north:	Camelopardus
west:	Auriga
south:	Gemini and Cancer
east:	Ursa Major.

α Lyncis, on a line from Regulus (α Leonis) and ε Leonis and as far beyond ϵ as ϵ is from α, is mag. 3. All the remaining stars are mag. 4 or fainter.

History

One of the constellations added by Hevelius in about 1690. It is said that he named it the Lynx because none but a lynx-eyed observer would be able to spot it.

APPENDIX 3

CANCER: The Crab

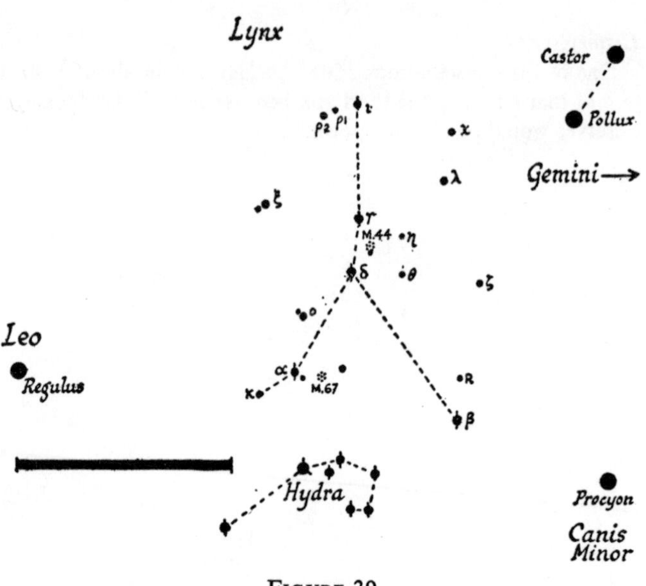

FIGURE 39

Position

A winter constellation culminating at midnight at the end of January. It lies between the conspicuous constellations Leo and Gemini, being bounded on the north by Lynx and on the south by Canis Major and the head of Hydra.

Description

A small constellation, easily overlooked unless its position between Gemini and Leo is known. It contains no star brighter than mag. 4. The brighter stars lie in the form of a tripod viewed from above. Its chief claim to distinction is the marvellous star cluster (see below) which it contains. The Crab is the 4th sign of the zodiac.

Mythology and History

In Greek mythology this is the crab which grasped the foot of Hercules while he was fighting the water snake in the Lernaean Marsh, translated to the skies by Juno (see under Hydra). Both the Chaldeans and the Platonists believed that Cancer was the 'Gate of Men' through which souls descend to earth to inhabit the bodies of new-born babies. The Romans consecrated the constellation to the god Mercury; the Egyptians, to their god Anubis.

The Praesepe (the cluster already referred to) was well known to the ancients, its earliest mention being by Hipparchus; the Greeks and Arabs likened it to a Manger (the stars γ and δ, η and θ, on either side of the cluster, being the Asses). It is now commonly known as the Beehive. Pliny notes that the visibility or otherwise of the Praesepe when the sky is clear is a reliable weather glass.

Objects of Interest

ζ Cancri. Interesting quadruple system, unfortunately not visible as such in small instruments. Of the four components, D (dark) revolves about C in under 20 years, C about A in approximately 50 years, and A and B revolve around each other in about 600 years. It is interesting to try and envisage this celestial whirligig while looking at the star.

ι Cancri. Test object for binoculars. Mags. 4·4 golden, 6·6 bluish-green; separation, 31″. Discovered by Herschel in 1782.

M. 44, N.G.C. 2632. The Praesepe, or Beehive Cluster. Visible to the naked eye as a misty spot. 500–600 stars (80 are visible with good glasses or the smallest telescope), yellowish and generally similar to the sun. Ideal object for binoculars. Distance, 520 L.Y.; diameter about 13 L.Y.

M. 67, N.G.C. 2682. Fine galactic cluster of some 500 stars (mags. 9–15), just visible as a nebulous spot in binoculars. Diameter about 13 L.Y.; distance 2,700 L.Y.

R Cancri. L.P.V., intermittently visible with binoculars. Mag. 6·2–11·2; period, 362d.

APPENDIX 3

CAMELOPARDUS: THE GIRAFFE

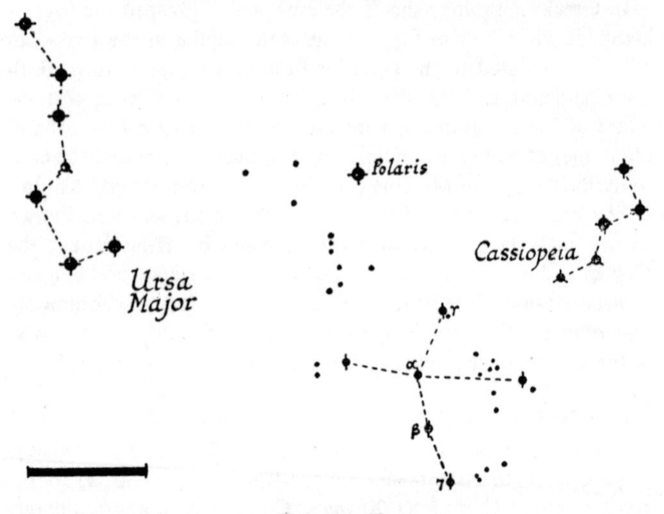

FIGURE 40

Position and Description

A large though inconspicuous circumpolar constellation lying between Cassiopeia and Ursa Major, bounded on the south by Lynx, Auriga and Perseus. Only six of the 150 stars visible to the naked eye are brighter than mag. 5. It contains nothing of interest for the observer with binoculars.

History

Camelopardus is a comparatively recent constellation, having been delineated by Bartschius in 1614.

LEO: THE LION

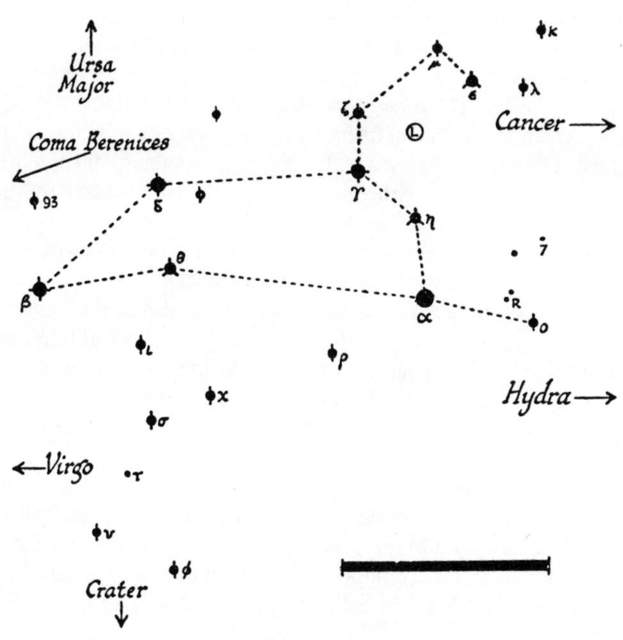

FIGURE 41

Position

A spring and early summer constellation, culminating at midnight, about March 1st; it is visible in the midnight sky from December to June. It is the first conspicuous grouping of stars eastward from Gemini, though actually the faint constellation of Cancer lies between them. To the north lies Ursa Major; to the south, Sextans and Crater; to the east, Coma Berenices and Virgo.

Description

A well defined and conspicuous constellation, Leo is the fifth sign of the zodiac. It bears a closer resemblance to the creature whose name it bears than do the majority of the constellations, the raised head and arched neck, forepaws and crouched body being fairly convincingly outlined by the stars. α Leonis (Regulus) is a fine 1st mag. star.

Mythology and History

Leo is among the most ancient of the constellations and has been known by this name from the earliest times of which we have record. The early Babylonians, Indians and all subsequent civilizations associated Leo with the sun. Leo was worshipped by the Egyptians and it is possible that the Sphinx is a representation of it. The national emblem of the ancient Persians, and their favourite architectural decorative device, was a crouched lion bearing the sun on its back; the royal standard of the United Kingdom merely carries on a very ancient tradition. A myth from classical times relates how Hercules slaughtered the terrible lion which haunted the forest of Nemea, Jupiter afterwards placing it among the stars to commemorate the feat.

Objects of Interest

α Leonis. Regulus. Wide double: mags. 1·3, 8·5; separation about 3′. Probably a binary, since the components have the same proper motion; a third component is shown by large instruments. Lies almost exactly on the ecliptic; distance, about 80 L.Y.

γ Leonis. Probably an optical pair. Primary a very fine binary (period, 407 years), not separable with binoculars. Distance, about 100 L.Y.; radial velocity, –25 m.p.s. Compare contrasted colours of γ and α by moving the eye quickly back and forth between them.

ε Leonis. Charming sight in binoculars, two mag. 7 stars forming a small triangle with ε.

ζ Leonis. Two, possibly three, companions to ζ are shown by binoculars.

Three faint doubles are:

τ Leonis. Mags. 5·5 yellowish-white, 7 pale blue; 90″.

7 Leonis. Mags. 6, 8; 42″.

93 Leonis. Mags. 4·7, 8·4; 74″.

R Leonis. Deep red L.P.V., partially visible with naked eye and binoculars. Mag. 5–10·5; period, 310d. ½° to the N.W. lies 18 Leonis (mag. 6) with which it must not be confused.

Leonids. Associated with Temple's comet. Once an important shower, giving particularly rich displays every 33 years (those of 1799, 1833 and 1866 being outstanding). Now weak, though there were good returns in 1932 and 1933. Maximum, about November 16th.

APPENDIX 3

URSA MAJOR: The Great Bear

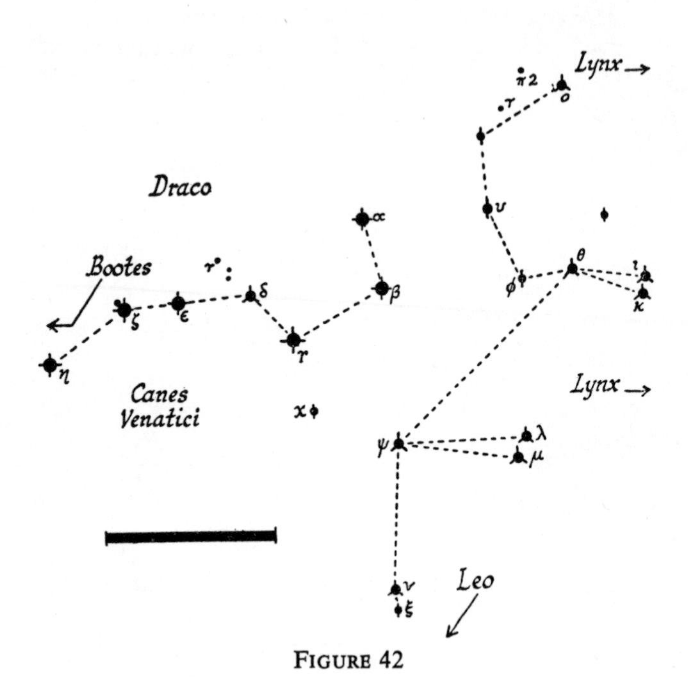

Figure 42

Position and Description

Ursa Major is circumpolar—i.e. it never sets—its distance from the north celestial pole being less than the altitude of the pole above the northern horizon. It is therefore of great value as a starting point from which the beginner can find other constellations (see pp. 21–22). The group of seven stars known as the Plough only occupies part of the area of sky covered by Ursa Major, though elsewhere there are no conspicuous star groups such as that formed by α, β, γ, δ, ϵ, ζ and η. The constellation is bounded on the north by Draco, which passes just north of the Plough in this region; its other boundaries are the constellations Bootes (if the curve of stars ϵ, ζ and η Ursae Majoris is continued, it strikes Arcturus), Canes Venatici (the northern part of which comes within a few degrees of the underside of the Plough's handle), Leo to the south-west, Lynx and Camelopardus.

Mythology and History

Ursa Major—or at any rate the seven stars of the Plough—is the oldest of the constellations, and was probably recognized and named by Man's remotest ancestors. It was called the Bear—the oldest known name for the constellation—by the Euphrates astronomers several millennia B.C., and it is an extraordinary fact, indicating the extreme antiquity of this designation and the persistence of folk memory, that it was thus known in the most widely separated parts of the world: Chaldeans, Persians, Indians, Phoenicians, Egyptians and North American Indians all knew the constellation as the Bear. The latter saw the four stars α, β, γ and δ as the Bear, the three stars of the handle being its hunters. The first carried a bow and arrow, the second a pot in which to cook the meal when the bear was killed, and the third an urn of stars to provide the fire for the pot. Each autumn the hunters succeeded in wounding the bear, and its blood dripping down upon the forests accounts for the reddening of the leaves during the fall. In one classical myth the Bear represented the beautiful nymph Callisto, who was loved by Jupiter. To save her from vengeance at the hands of the jealous Juno he changed her into a bear, but Juno prevailed on Diana, goddess of the chase, to kill the bear. Jupiter translated her to the skies.

The Egyptians did not call Ursa Major the Bear but the Haunch. The name Wain is common to all Teutonic peoples, probably originally associated with Odin as Orion with his consort Frigga or with Thor.

APPENDIX 3

Alternative names for the Plough are the Dipper, and Charles' Wain. The stars β and α are universally known as the Pointers since they point at, or rather very near, Polaris.

Objects of Interest

ζ Ursae Majoris. Mizar. Forms with Alcor (mag. 5, distant 11′ 30″) an easy naked-eye pair. The Arabs, who were exceptionally keen-sighted observers, considered this pair a severe test of eye-sight, and it has therefore been suggested that the appearance of the pair may have changed in some way during the past 1000-odd years. It is interesting to reflect, while looking at Mizar and Alcor, that light takes 90d. to cross the gap between them (c.f. 8 m. from sun to earth, 4½ years from sun to nearest known star). Mizar it-self is a telescopic binary, discovered by Riccioli, but the separa-tion of 14″ forbids its observation with binoculars; distance about 80 L.Y. It was the first double to be photographed, while the comes was the first sp. bin. to be discovered (1889): period, 20d. 13h.; mean distance apart, about 1·7 light minutes. Binoculars show several faint stars, about mag. 8, in the vicinity of the Mizar-Alcor system.

T Ursae Majoris. L.P.V. (254d.); mag. 5·5–13·6. Therefore visible to the naked eye at maximum and invisible with binocu-lars at minimum.

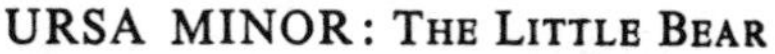

URSA MINOR: The Little Bear

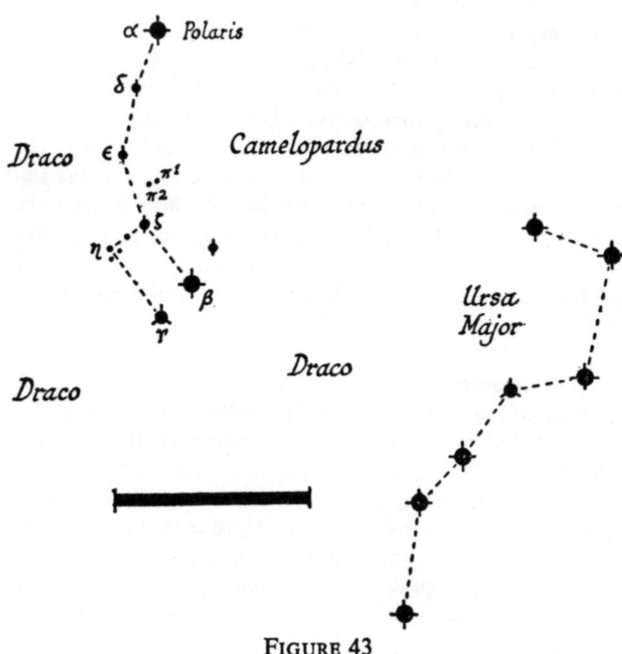

Figure 43

Position

Ursa Minor is, like Ursa Major, circumpolar. It is easily located between Draco and Polaris, to the north of the Plough.

Description

A small constellation, possessing two 2nd and one 3rd mag. stars. It is chiefly notable for the fact that it includes the Pole Star within its boundaries: Polaris is α Ursae Minoris. The stars β and γ are often known as the Guardians of the Pole.

Mythology and History

In Greece this constellation owed its origin to Thales, *c.* 600

B.C. To the Arabs it was the Lesser Bear, to the Phoenicians the Guider. Among northern countries, the Finns also called it the Lesser Bear, while the Danes knew it as the Smaller Chariot. Various classical myths refer to the constellation. One, a variant of that already quoted under Ursa Major, tells how Callisto was turned into a bear by Juno herself, and when Callisto's son Arcas was on the point of killing her in the chase Jupiter turned him into a bear also, and placed both mother and son among the stars. Another myth claims that the two Bears are those which nursed Jupiter upon Mount Ida. Still another supposes that the Bears were placed among the constellations in reward for hiding Jupiter from his father Kronos. Ursa Minor is often known as the Little Dipper.

Objects of Interest

α Ursae Minoris. Polaris is situated rather more than 1° from the true celestial pole, and therefore describes a small diurnal circle upon the star sphere. Long exposure photographs have shown as many as 200 faint stars within this small diurnal circle, 2½° in diameter. It will be nearest the pole (25′) towards the end of next century. Mag. 2·0; brightness, 2,500☉; distance, 650 L.Y.; radial velocity, –15 m.p.s. Polaris is a sp. bin.; period, 3·97d. Good glasses will show several faint stars near by (one of mag. 7), but not the mag. 9 companion, which is a test object for a 3-in. telescope.

π^1 Ursae Minoris. The faint comes may just be glimpsed with good, firmly mounted binoculars. Mags. 6, 7; separation, 30″.

Ursids. Maximum (December 22nd) of short duration. Major display on the occasion of its discovery in 1945; since then, 10–20 per hour at maximum. Possibly connected with comet Méchain-Tuttle. The Radiant is close to *β*.

CORVUS AND CRATER

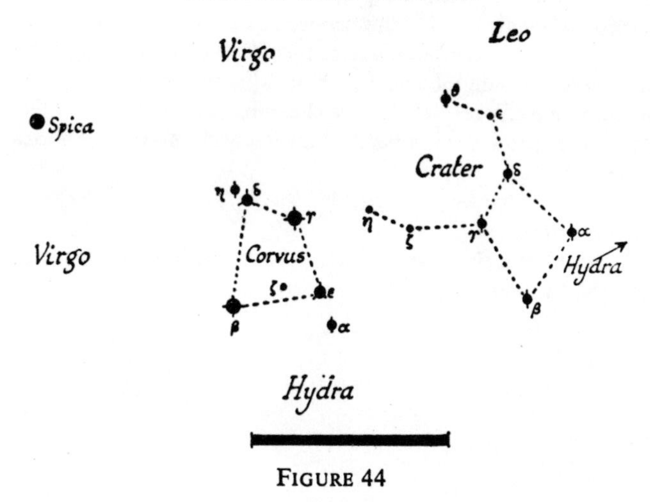

FIGURE 44

CORVUS: THE CROW

Position

Corvus lies 20° south of the celestial equator between Virgo and Hydra. It culminates at midnight towards the end of March, when it is best observed. It is easily found some 18° south-west of Spica (α Virginis).

Description

A small constellation, the four brightest stars of which (mags. 2, 3) form a quadrilateral. α, mag. 4, lies 2° south of the south-west corner of the quadrilateral.

Mythology and History

Known as the Raven or Crow by the Greeks, Romans and Hebrews, while the ancient Chinese astronomers called it the Red Bird. Several myths associate the constellation with a raven or crow. According to the Greek legend, Apollo fell in love with

Coronis and, being a jealous lover, sent a crow to spy on her. Coronis was unfaithful to Apollo, which lapse the crow duly reported. In repayment for this service Apollo gave it immortality among the stars. Another tradition links Corvus with the daughter of Coronaeus, king of Phocis, whom Minerva rescued from the amorous pursuit of Neptune by changing her into a crow. Old star atlases show Corvus perched on the back of Hydra, the sea serpent.

Objects of Interest

ζ Corvi. Binoculars show a faint companion. Radial velocity of primary, –13 m.p.s.

CRATER: The Cup

Position and Description

A small and undistinguished constellation lying due west of Corvus, which like the latter is figured in old star atlases as standing upon the coils of the sea serpent. α, δ, γ and β form a quadrilateral of mag. 4 stars, like a fainter twin or mirror image of Corvus, 15° to the east. The constellation contains no objects of specific interest for observation with binoculars, but provides some attractive sweeping.

APPENDIX 3

HYDRA: THE SEA SERPENT

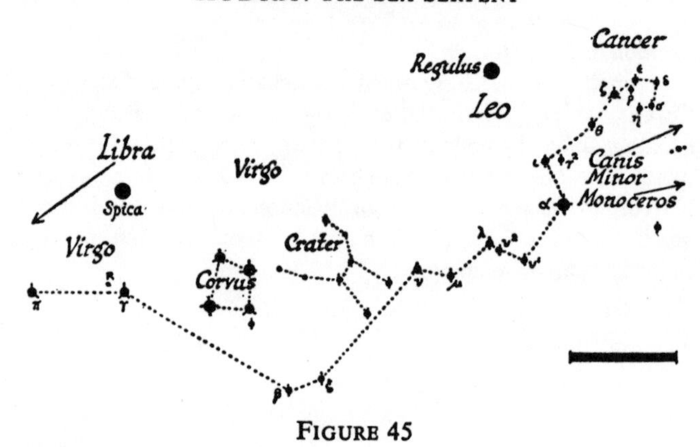

FIGURE 45

Position

Hydra is a long constellation, stretching for some 100° between its eastern and western boundaries; the greater part of it lies south of the celestial equator. In the west it is bounded by Cancer, Canis Minor and Monoceros. Thence it runs south and east below Sextans (a small constellation south of Leo), Crater, Corvus, Virgo and Libra. The west end of the constellation culminates at midnight at the end of January; the east end during April.

Description

The Sea Serpent, or Water Snake, is not well placed for observation in our latitudes owing to its southern declination. The head of Hydra, some 15° due east of Canis Minor, is marked by a group of 3rd and 4th mag. stars, easily identified. Thence the coils of the serpent drop south towards the 2nd mag. α; the section south of Crater, Corvus and Virgo is quite undistinguished. Towards the tail are the two 3rd mag. stars γ and π.

Mythology and History

To the Egyptians this long, wandering constellation was the celestial counterpart of the Nile, but the Arabs followed the

Greeks in naming it the Serpent. According to Greek mythology this was the water snake that inhabited the Lernaean Marshes, whence it ravaged the province of Argos. It had nine heads, and as soon as one was cut off another would sprout in its place; this made it difficult to kill. Hercules finally succeeded in exterminating the monster with the aid of Iolas, who branded each severed neck with a hot iron to prevent the new head growing. The ninth head, said to be immortal, Hercules buried beneath a rock. During the course of the struggle Juno, envious of Hercules' fame and prowess, sent a crab to distract his attention by nipping his foot; this plan failed, however, and the crab was quickly despatched (see under Cancer).

Objects of Interest

α Hydrae. Notable on account of its isolated position and its orange-red colour.

R Hydrae. L.P.V. (415d.); mag. 4–10.

APPENDIX 3

COMA BERENICES: BERENICE'S HAIR

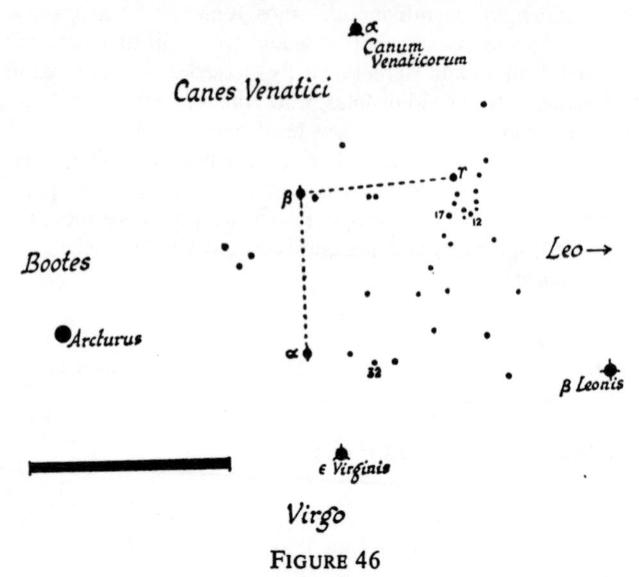

FIGURE 46

Position

A spring and summer constellation culminating at midnight early in April. It lies between Bootes and Leo and is bounded on the north by Canes Venatici, on the south by Virgo. It lies roughly within the triangle marked by α Bootis (Arcturus), α Canum Venaticorum and β Leonis (Denebola).

Description

A small constellation, unremarkable to the naked eye but full of interest when swept over with binoculars or a small telescope. α and β, the brightest stars in Coma, are mag. 4. To the naked eye it appears as a large cloud of faint stars, thirty or more being clearly seen with binoculars.

Mythology and History

Berenice was the beautiful young wife of Ptolemy Evergetes,

king of Egypt. When her husband left her for the wars she swore to the gods that if he returned safely she would sacrifice her hair, famed for its loveliness, to Venus the goddess of beauty. He did return safely, and Berenice in gratitude fulfilled her promise. But the following day it was discovered that the tresses had disappeared from the temple, the guardians of which were on the point of being executed when Conon, a Greek astronomer, came to the king and told him that Venus had translated the missing hair to the sky. In support of his contention he pointed to the glittering cloud near Arcturus, and the king, knowing little of astronomy, believed; the temple guardians were released.

Objects of Interest

12 Comae. Mags. 4·5, 8, yellow and bluish; separation, 66″.

17 Comae. A wide double: mags. 4·8, 6; separation, 145″.

32 Comae. A wide and faint double: mags. 5·6, 6; 195″.

APPENDIX 3

VIRGO: THE VIRGIN

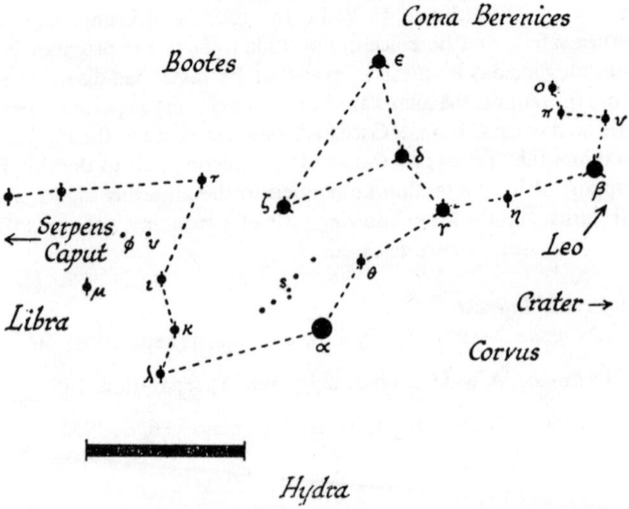

Coma Berenices
Bootes
Serpens Caput
Libra
Leo
Crater
Corvus
Hydra

FIGURE 47

Position

Virgo is visible throughout the summer months, culminating at midnight early in April. Its boundaries are:

West:	Leo
South-west:	Crater and Corvus
South:	Hydra
East:	Libra and Serpens Caput
North:	Bootes and Coma Berenices

Spica (α Virginis) forms the apex of an inverted equilateral triangle whose base angles are marked by Denebola (β Leonis) and Arcturus (α Bootis). It also marks the lower base angle of an isosceles triangle whose other corners are Arcturus and Regulus (α Leonis). The celestial equator passes through Virgo close to ζ and η.

Description

Spica is a fine first magnitude star of the purest white tint, lying in the south of the constellation. Virgo possesses in addition five 3rd mag. stars, so that it is a sufficiently conspicuous grouping. On its northern border is situated a large clustering of faint extra-galactic nebulae, unfortunately not to be seen with binoculars.

Mythology and History

Virgo is the sixth sign of the zodiac, and a constellation of great antiquity. In India it was called the Maiden, and in the Euphrates area was identified with Ishtar (= Venus), the queen of the heavens. Two ancient myths referring to Virgo are worth noticing. The first associates the constellation with Astrae, goddess of Justice during the Golden Age when the gods lived among men on earth. The increasing iniquity of the human species, however, drove them one by one to forsake the earth for heaven. Astraea was the last to leave, taking with her the scales of justice which are still to be seen by her side in the sky. An Egyptian tradition relates how the goddess Isis, while fleeing the monster Typhon, in her haste dropped a sheaf of corn she was carrying. Isis herself was translated to the skies as Virgo, while the scattered grain from her sheaf became the Milky Way.

Objects of Interest

α Virginis. Spica, mag. 1. Clear white colour (compare with the red δ); brightness, 1,500☉; distance, 212 L.Y.

S Virginis. L.P.V. (372d.). Mag. 5·6–12·5.

SS Virginis. A variable (mag. 6–9) notable for its deep red tint.

APPENDIX 3

CANES VENATICI: THE HOUNDS

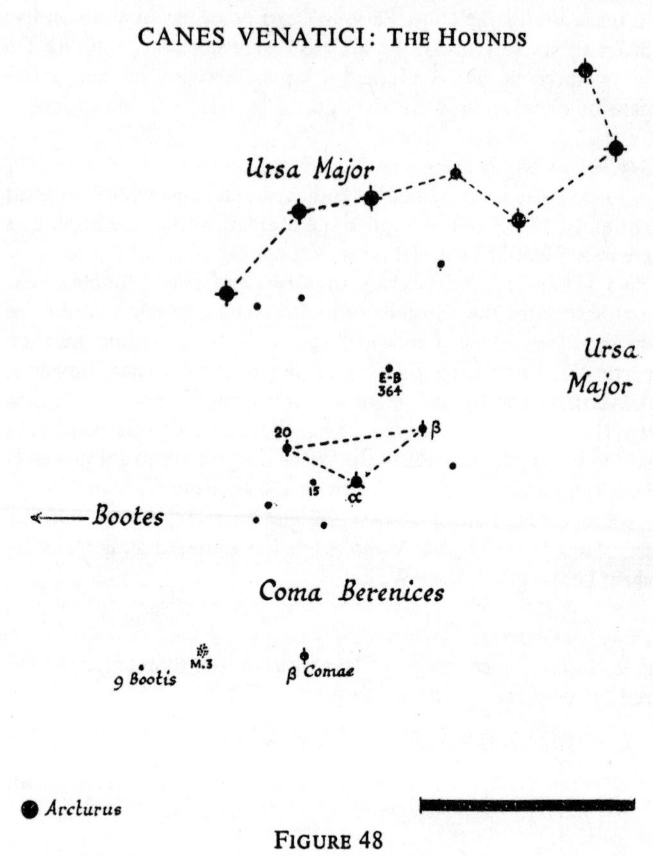

FIGURE 48

Position

Due south of the handle of the Plough. Bounded on the east by Bootes, the west by the southern reaches of Ursa Major, and on the south by Coma Berenices. The constellation is visible for the greater part of the year, culminating at midnight on April 7th.

Description

The only conspicuous star in this small constellation is Cor

Caroli, mag. 3·2 (α). The area between the handle of the Plough and Coma Berenices is well scattered with faint stars.

Mythology and History

The two Hounds are pictured in old atlases as being held in leash by Bootes, hunter of the Bear. It is, compared with most, a recent constellation, having been delineated by Hevelius in about 1690. The hounds are named Asterion and Chara, α being located in the latter. Cor Caroli (Charles's Heart) was so named by Halley, Astronomer Royal in the time of Charles I.

Objects of Interest

15 Canum Venaticorum. Test object for binoculars. Mags. 5·5, 6; separation, 290″.

E–B 364. Mag. 5 star worth finding for its brilliant red colour.

M. 3, N.G.C. 5272. One of the finest globular clusters in the northern skies; just visible to the naked eye on clear moonless nights, 6° east of β Comae. Some 30,000 stars of mags. 11–16, including over 130 Cepheid-type variables with periods ranging from 11–18 hrs., average stellar mag. 15. Distance, nearly 45,000 L.Y.; diameter 7′, or 490 L.Y. Radial velocity, –78 m.p.s. In binoculars it appears as a circular nebulosity with no trace of stellar resolution.

APPENDIX 3

BOOTES: THE HERDSMAN

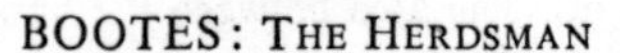

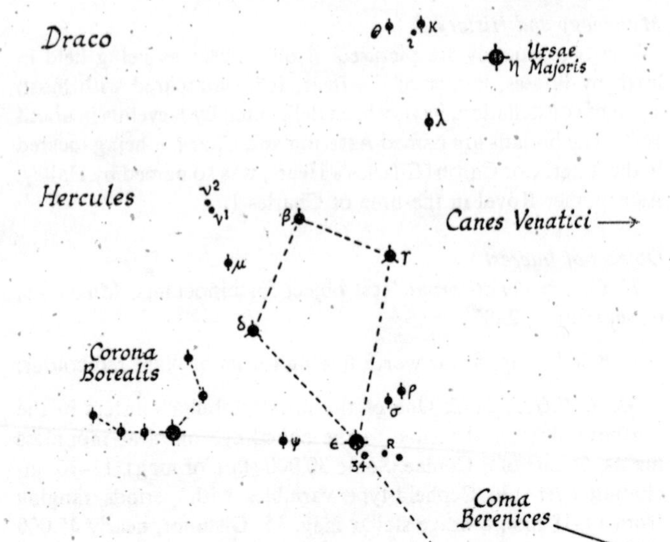

FIGURE 49

Position

Summer constellation, culminating at midnight at the beginning of May. To its east lie Serpens and the easily spotted constellation of the Crown; to its west lie Canes Venatici and Coma Berenices; to the south and south-west, Virgo.

Description

A fairly conspicuous constellation containing the 1st mag. star

Arcturus. Its brighter stars—α, ϵ, δ, β and γ—mark the outline of a great kite flying the dark summer skies. In its northern corner Bootes abuts upon the handle of the Plough, and in this region, marked by the star θ Bootis, there is some good low power sweeping suitable for binoculars. Arcturus is one of the four bright stars forming the so-called 'Diamond of Virgo', a feature of the night skies of summer; the other corners are marked by α Virginis (mag. 1·2), β Leonis (2·2) and α Canum Venaticorum (3·2).

Mythology

Bootes is the Ploughman, eternally driving the Plough around the north celestial pole; in this identification he is Arcas, son of Jupiter and the nymph Callisto, and inventor of the plough. In some traditions he figures as a huntsman with two dogs on leash (Canes Venatici), tracking down the Bear. A later but now obsolete name for the constellation is Atlas, since Bootes, from its proximity to the pole, appears when on the meridian to be holding up the heavens. The constellation is mentioned by Homer, while Arcturus is mentioned in the Book of Job, for which reason it is sometimes known as 'Job's star', though there is some doubt as to the validity of this identification.

Objects of Interest

α *Bootis*. Arcturus, mag. 0·2, notable for its deep yellow tint. Distance, 36 L.Y.; luminosity, 115☉; diameter, 25☉. The heat received from Arcturus has been measured directly by means of sensitive thermocouples used in conjunction with a telescope of large aperture, and has been found to be about the same as that received from a candle 5 miles away—a striking indication of the precision and sensitivity of the instruments that the modern astronomer has at his disposal. Arcturus has an unusually large proper motion, first noticed by Halley: 85 m.p.s., 3′ 50″ per century, or the apparent diameter of the moon since the time of Ptolemy. Sp. bin., period 212d. Binoculars reveal several faint stars dotted round it.

δ *Bootis*. A difficult double for binoculars. Mag. 3 yellow, mag. 7·4 blue; separation, 105″.

ι Bootis. Mags. 5, 7·5, separation 38″. The two components have a common proper motion and are therefore probably a true binary. The primary is itself a close double.

μ^1, μ^2 *Bootis.* Mags. 4, 6·7; separation, 108″. μ^2 is itself a binary. period, about 230 years.

R Bootis. Varies between mags. 5·9 and 13·0 in 222d. With binoculars it is observable for nearly half of its period.

ζ Bootids. March 10th–12th; swift moving. Weak shower.

DRACO: THE DRAGON

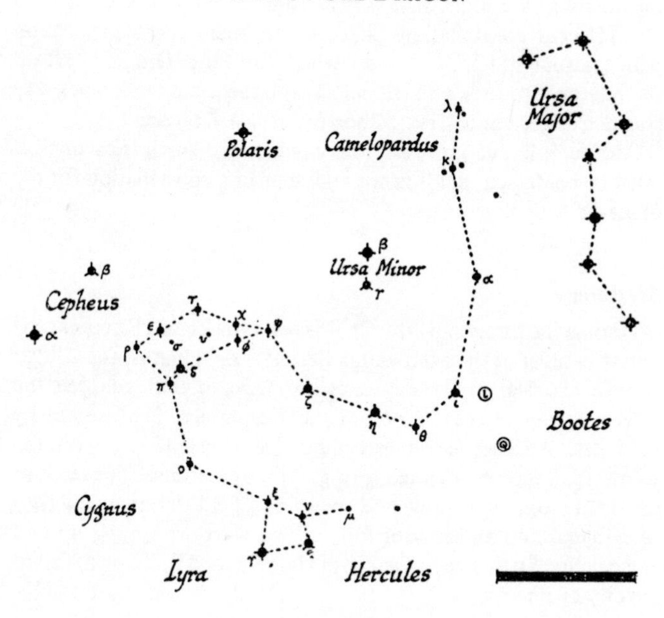

FIGURE 50

Position and Description

A circumpolar constellation lying between Polaris and the northern horizon at midnight during the winter months, and 'above' (or south of) Polaris during the summer. It culminates at midnight throughout April and May. Its position in relation to neighbouring constellations is best seen from the accompanying map, which shows the winter aspect of the northern sky with Draco between Polaris and the northern horizon.

Owing to its size (it stretches half-way round the north celestial pole) and straggling shape (it is bounded by no less than eight constellations) and also to the comparative faintness of its stars (none brighter than the 2nd mag.), Draco is not too easy to trace immediately in the night sky. It is as well to acquaint oneself with the adjacent constellations first and to begin the identification with the four stars β, γ, ξ and ν which form the 'Dragon's Head' (by reference to Lyra and Hercules); and thence to work one's way round Cepheus and Ursa Minor towards λ Draconis.

Although Draco possesses many objects for small telescopes, it must be confessed that it is not an interesting constellation for the observer with binoculars.

Mythology

Named the Dragon by the Chaldeans, Greeks and Romans; the Persians called it the Man-eating Serpent; the Hindus, the Alligator. In classical mythology it is the Dragon that guarded the golden apples in the Garden of the Hesperides, finally slain by Hercules. Another story associates the constellation with the guardian dragon of a sacred spring, slain by Cadmus. The teeth of the dead monster he drew and sowed in a field. Immediately they germinated and an army of fully armed warriors sprang up and engaged in battle. The five survivors later helped Cadmus to found the city of Boeotia.

Objects of Interest

ν Draconis. Inconspicuous double, separable with binoculars. Mags. 4·5, 4·5; both white; separation, 62″. The two components have the same proper motion.

ι Draconids. Once an important shower (maximum, June 29th), but now extinct. Connected with the Pons-Winnecke comet, period 6·1 years.

Giacobinids. Radiant close to β. Short maximum on October 10th. Discovered 1926. Associated with Giacobini's comet, period 6·6 years. Good displays in 1933, 1946, 1952.

Quadrantids. About 40 per hour may be expected at maximum (January 4th). For observers in the British Isles, the radiant (near the border of Draco and Bootes) is circumpolar, as are those of the other Draconid showers.

APPENDIX 3

LIBRA: THE SCALES

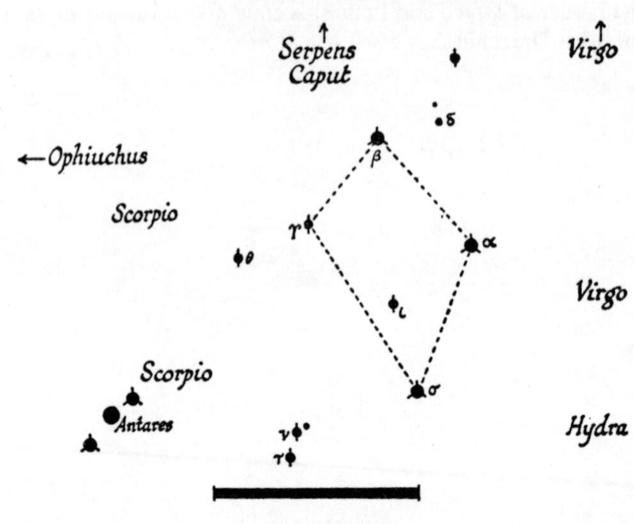

FIGURE 51

Position

A spring constellation, culminating at midnight early in May. It lies just below the celestial equator between Scorpius and Ophiuchus on the east, and Virgo on the west; it is bounded on the north by Serpens Caput. The triplet of bright stars in Scorpius—τ, Antares and σ, a twin of Aquila—points between α and β Librae.

Description

Libra is the seventh sign of the zodiac. It is not a noticeable constellation, but the kite-like figure outlined by its four brightest stars is easily spotted.

Mythology and History

The sun is in Libra at the autumn equinox, when the days and nights are of equal length; this is supposed to be the reason for identifying the constellation with Scales. The name Libra is due to the Romans. The Greeks regarded it as part of Scorpius, not as a separate constellation, but in more ancient times it had been represented by a figure holding a pair of scales: Indians, Chinese, Egyptians and Hebrews all saw the constellation thus. It is sometimes said that Libra commemorates Mochis, traditionally the inventor of weights and measures.

Objects of Interest

α Librae. Mags. 3, 5·3; separation, 230″. Separable by acute naked eyesight.

β Librae. In general, green or blue stars are all faint, usually the comites of binaries; β Librae is the only star visible to the naked eye as being distinctly green. It has a radial velocity of –6 m.p.s.

δ Librae. Algol-type eclipsing variable. Mag. 4·9–6·2, and therefore well observable with binoculars; period, 2d. 7h. 51m. Discovered as recently as 1859.

APPENDIX 3

CORONA BOREALIS: The Northern Crown

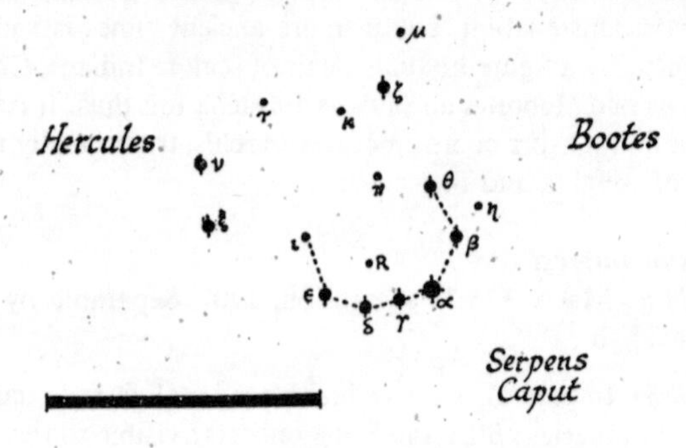

FIGURE 52

Position

A spring and summer constellation, culminating at midnight about May 20th. It lies between Bootes and Hercules, which constellations also border it on the north; to the south lies Serpens Caput.

Description

The Northern Crown is a small constellation, but quite unmistakable. If the eye is carried some 20° north-east of Arcturus it will notice a semicircle of seven stars: one (α) mag. 2, the remainder mag. 4. Further to the north and east lie ζ, ν, and ξ, all of the 4th magnitude.

Mythology and History

The Greeks, Romans and Hebrews all knew this constellation as the Crown or Wreath, and as early as 500 B.C. it was believed to be the crown which Bacchus gave Ariadne, daughter of King Minos of Crete, to comfort her after she had been deserted by Theseus.

Objects of Interest

T Coronae. A most interesting star, having some of the features of a variable and some of those of a nova. At present too faint to be found with binoculars—mag. 9·5. In May 1866 it rose suddenly from mag. 9·5 to 2·2. Within nine days it was again invisible to the naked eye; after sinking to mag. 9 several weeks later it rose again to mag. 7. Finally, it relapsed to its present magnitude. There are several unusual features in this outburst: T Coronae is much further from the Milky Way than is usual with novae; it was brighter before the outburst than most novae; its period of maximum brightness was unusually brief. It was the first nova to be studied spectroscopically—by Huggins, one of the pioneers of the use of this instrument in astronomy.

R Coronae. Irregular variable; mag. 5·8–12·5. Usually remains at maximum for several years; the fall to minimum is relatively rapid, and the duration of minimum varies within wide limits. The brighter section of the light curve is visible with binoculars.

ν Coronae. Wide double. Mags. 4·8, 5, yellow; separation 370″.

APPENDIX 3

SCORPIUS: THE SCORPION

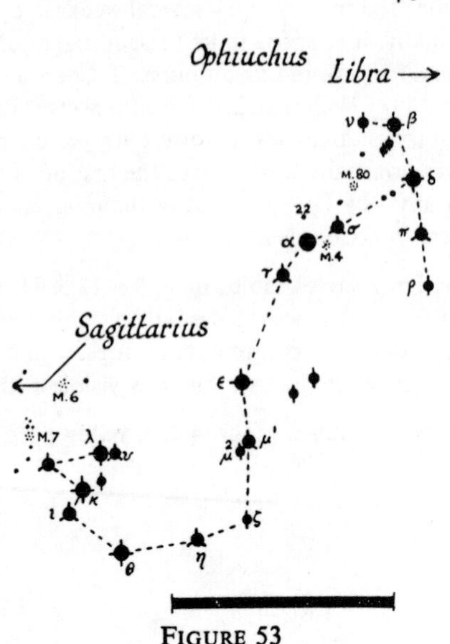

FIGURE 53

Position

A summer constellation, culminating at midnight early in June. It lies entirely south of the celestial equator, and should therefore be observed when on or near the meridian. The visible part of the constellation (for the southern reaches of Scorpius never rise above the horizon in our latitudes) lies between Ophiuchus on the east and Libra on the west. α Scorpii (Antares) is a conspicuous first magnitude star, and attracts the eye to the constellation.

Description

Scorpius lies in the richest region of the Milky Way, and it is in this direction that the centre of the Galaxy or Stellar System lies. The northern area of the constellation is characterized by a conspicuous line of bright stars: α (mag. 1), β, δ, ε (mag. 2), and τ, σ,

π (mag. 3). The three stars α, τ and σ form a slanting line, reminiscent of the most conspicuous feature of the constellation Aquila. There is extremely fine sweeping in many of the regions of Scorpius, especially southward from β. It is a misfortune that this fascinating constellation lies so low in our northern skies.

Mythology and History

Old star atlases show Sagittarius (the Archer, *q.v.*) aiming his arrow at Antares, the heart of the Scorpion. The myth associating Scorpius with Orion (*q.v.*) has already been related. It has been suggested that this sign of the zodiac was named the Scorpion for the reason that sickness and the plague were particularly rife in Egypt during the month when the sun is in the eighth sign. Greeks, Romans, Persians and Arabs all knew the constellation by this name. In China at the time of Confucius it was called the Great Fire, referring to the fiery red glow of Antares. Antares means 'rival of Mars', and some 3,000 years before Christ it was one of the four 'Royal Stars' of Persian astronomy. It is on record by both Greek and Chinese astronomers that in the year 134 B.C. a nova was observed in Scorpius. Other early records refer to novae which appeared in this constellation in A.D. 393, 827, 1203 and 1578.

Objects of Interest

α *Scorpii.* Antares. Brilliant mag. 1 red star, and one of the largest yet measured; diameter, 370,000,000 miles, or about 430 ☉. Were it transported to the position now occupied by the sun, the orbits of Mercury, Venus, the Earth, Mars and many of the asteroids would lie below its surface. Brightness, about 3,000 ☉; hence, area for area, its luminosity is only 0·01 ☉. It has a minute, green companion 3″ distant; it is a true binary. Distance in the neighbourhood of 425 L.Y. Binoculars show several faint companions.

μ *Scorpii.* Naked-eye double. Mags. 3, 4.

ν *Scorpii.* Test object for binoculars. Mags. 4·2, 6·5; separation, 41″. Both are close binaries.

λ *Scorpii.* Guide star to the clusters M. 6 and M. 7, both visible in binoculars; its luminosity is about 2,000⊙, and its distance 290 L.Y.

ω *Scorpii.* Two mag. 4·5 stars, well placed in relation to β.

22 Scorpii. Mag. 5, with two faint companions; easily found from its proximity with Antares.

M. 80, N.G.C. 6093. A fine cluster, situated about midway between α and β, and appearing in binoculars as a circular nebula 5½′ in diameter; mag. 7·8. The variable T Scorpii appears to be centrally placed within the cluster but is probably very much nearer to the sun. M. 80 contains thousands of stars, average mag. 14. Distance, about 65,000 L.Y. Beautiful neighbourhood.

M. 4, N.G.C. 6121. Cluster of faint stars, about 1½° W. of Antares. Diameter, 12′.

M. 6, N.G.C. 6405.
M. 7, N.G.C. 6475. } Fine open clusters, well seen in binoculars.

SERPENS: The Serpent

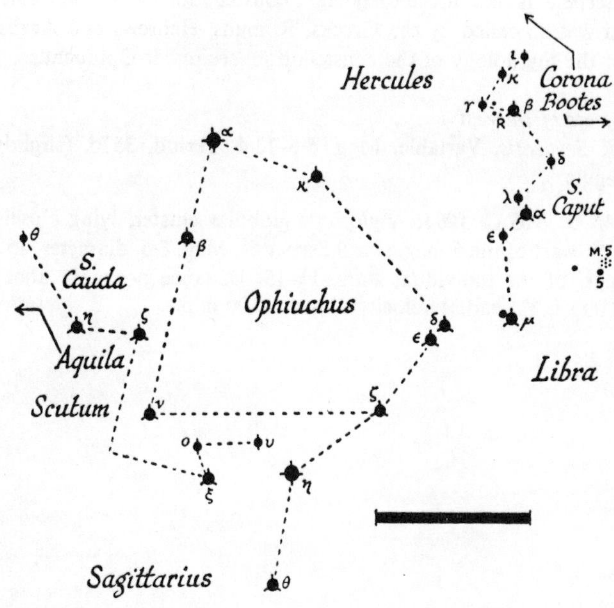

FIGURE 54

Position and Description

Serpens is unique among the constellations in consisting of two completely separated areas, with another constellation (Ophiuchus) between them. The two halves of the constellation—known as Serpens Caput (Head) and Serpens Cauda (Tail)—both lie on the celestial equator. Serpens Caput, the western of the two, is bounded by Corona on the north, Hercules and Ophiuchus on the east, Libra on the south, and Virgo and Bootes on the west; Serpens Cauda by Ophiuchus on the west, Ophiuchus and Aquila on the north, Aquila and Scutum on the east. Serpens is a spring and summer constellation, culminating at midnight during May and June. Caput contains the third magnitude stars α, β and μ; Cauda, η and ξ of the same magnitude.

History

Serpens is one of the forty-eight constellations of the ancients, and was so called by the Greeks, Romans, Hebrews and Arabs. For the mythology of the constellation, see under Ophiuchus.

Objects of Interest

R Serpentis. Variable. Mag. 5·5–13·4. Period, 357d. (slightly irregular).

M. 5, N.G.C. 5904. A glorious globular cluster, lying closely north-west of the 5 mag. star 5 Serpentis. Mag. 6·5, diameter, 15′. Mags. of the individual stars, 11–15. Distance probably about 30,000 L.Y.; radial velocity about −100 m.p.s.

OPHIUCHUS: THE SERPENT CARRIER

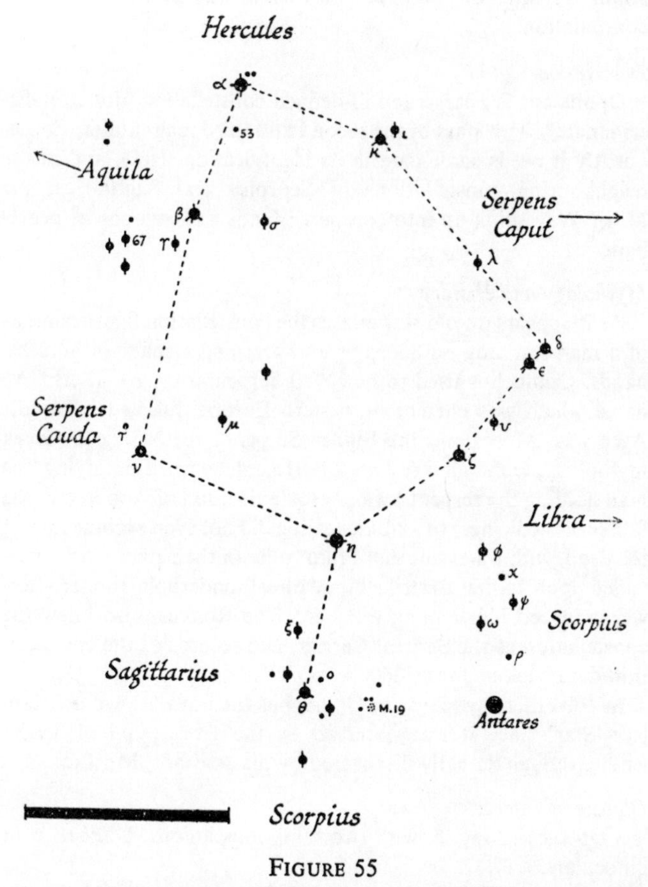

FIGURE 55

Position

A summer constellation, culminating at midnight about June 10th. The celestial equator cuts through Ophiuchus, which is bounded on the north by Hercules; on the west by Serpens Caput, Libra and Scorpius; on the east by Sagittarius, Serpens Cauda and Aquila. α Ophiuchi lies about 33° west and slightly north of Altair (α Aquilae)—perhaps the easiest way of identifying the constellation.

Description

Ophiuchus is a large and ill-defined constellation, dotted indiscriminately with stars of the second and third magnitudes. Consequently it needs some care in its identification. Here, and in the neighbouring constellations of Scorpius and Sagittarius, the Milky Way splits up into complex islands and lagoons of pearly light.

Mythology and History

As it appears on old star atlases the constellation figure consists of a man standing on Scorpius and grasping a snake in both his hands. Ophiuchus used to be called Serpentarius. An alternative name which was current in western Europe during the Middle Ages was 'Moses and the Brazen Serpent'. According to Greek mythology, Ophiuchus (a Greek portmanteau word signifying 'the man holding the serpent') was Aesculapius, son of Apollo and the father of medicine. His skill in healing did not even exclude raising the dead, with the result that Pluto, ruler of the nether world, prevailed upon Jupiter to strike him with a thunderbolt. Jupiter afterwards placed him among the stars. The Romans also knew the constellation as the Serpent Carrier, and records of the constellation date back as far as 2000 B.C.

In 1604 there appeared in Ophiuchus the nova known as 'Kepler's Star' since it was observed by the great pupil of Tycho Brahe, though actually discovered by his assistant, Mostlin.

Objects of Interest

ρ Ophiuchi. Mag. 5, with two faint companions. Difficult with binoculars.

53 Ophiuchi. Faint pair: mags. 5·6, 7·3; separation, 41″.

67 Ophiuchi. Mag. 4 yellow, mag. 8 reddish; separation, 54″.

M. 19, N.G.C. 6273. Cluster, just invisible to the naked eye but easily swept up with binoculars. Mag. 6·8, diameter about 5′. One distance estimate is 52,000 L.Y.

HERCULES

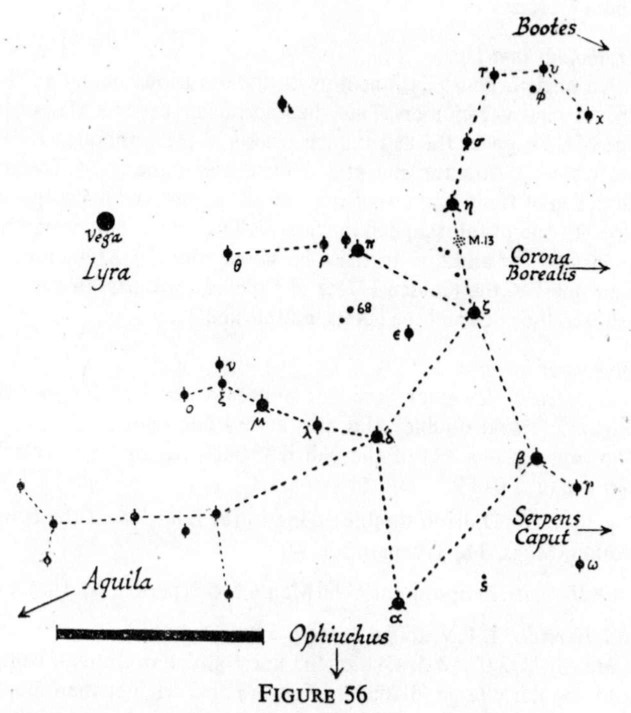

FIGURE 56

Position

A summer constellation, culminating at midnight about the middle of June. It is bounded by Draco on the north; Bootes, Corona and Serpens on the west; Ophiuchus on the south; and Lyra, Aquila, Sagitta and Vulpecula on the east.

Description

Hercules is a large constellation with stars of different magnitudes scattered all over it: unlike Orion, for instance, it does not consist of a few bright stars which form an obvious pattern, all the remainder being faint. It is therefore not an outstandingly conspicuous constellation; it is, however, quite easily found, lying for the most part within the triangle Vega (α Lyrae), Altair (α Aquilae) and α Coronae.

Mythology and History

An ancient constellation, known under various names to the pre-Grecian astronomers. Thus the Phoenicians called it Melkarth, one of their gods; the astronomer-priests of the Euphrates valley associated it with the sun god. Hercules was the great Theban hero, son of Jupiter and Alcmena, whose courage and phenomenal strength found full scope in the famous Twelve Labours which he undertook. In addition to these he sailed with the Argonauts in their quest of the Golden Fleece of Colchis, took part in the war between the gods and the giants, and sacked Troy.

Objects of Interest

α Herculis. Irregular variable; unusually red star for one so bright; a visual double; and also a very fine telescopic double. The brighter member of the pair is variable; period (irregular), 88d.; mag. 3·0–3·9.

γ Herculis. Difficult double owing to the faintness of the companion. Mags. 3·8, 8; separation, 40″.

68 Herculis. Eclipsing variable. Mag. 4·8–5·3; period, 2d. 1h. 12m.

S Herculis. L.P.V. (300d.); mag. 5·9–12·5.

M. 13. N.G.C. 6205. One of the finest globular clusters, being both angularly large (diameter about 6′) and brighter than most. Discovered by Halley, 1716. When its position is known, naked eyesight shows it as a mag. 6 star on moonless nights. A telescope is required to begin resolution of the individual stars, mags. 12–15. Photography shows that it contains some 30,000 stars brighter than mag. 21; total number probably in the neighbourhood of 100,000. Probable distance, 27,000 L.Y., corresponding to diameter of some 11 L.Y. Radial velocity about −150 m.p.s.

SCUTUM: The Shield

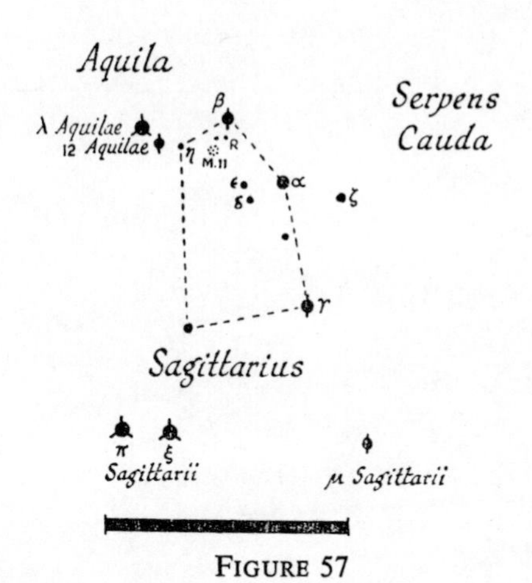

FIGURE 57

Position and Description

A small and undistinguished constellation (α is only mag. 4) lying in the Milky Way just south of the celestial equator. It is bounded by Sagittarius on the south, Aquila on the east and north, Serpens Cauda on the north and west. It culminates at midnight at the beginning of July. The indications given in the accompanying map enable it to be found without difficulty. β and γ are mag. 4·5, the remaining stars (apart from α) being fainter than this.

Historical

Created by Hevelius about 1690. Its full name is Scutum Sobieski.

Objects of Interest

R Scuti. Irregular variable, mag. 4·8–7·8.

M. 11, N.G.C. 6705. Fine cluster of the condensed galactic type, situated on the N. edge of one of the Sagittarius star clouds. May just be seen with the naked eye; nebulous spot in binoculars. Diameter about 12′ (20 L.Y.); distance, 5,500 L.Y.

LYRA: THE LYRE

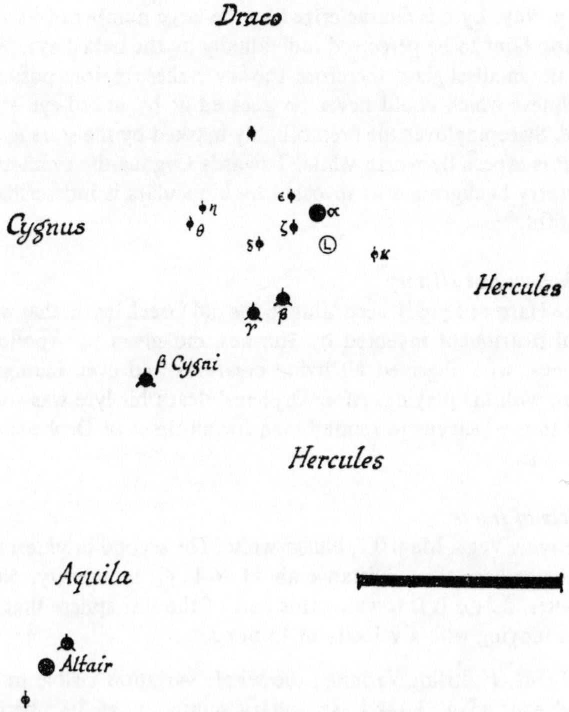

FIGURE 58

Position

A summer constellation, culminating at midnight at the beginning of July. It is situated between Cygnus and Hercules; its proximity with the former and the brilliance of Vega (α Lyrae) make it easy to find. The three bright stars of Aquila (Altair, β and γ) point straight at it.

Description

To the naked eye only three stars are prominent. These form an acute-angled triangle with Vega at its apex and the 3rd mag. γ and β at the corners. Like all the constellations situated in or near the Milky Way, Lyra is characterized by the large number of its stars just too faint to be perceived individually by the naked eye. With even the smallest glass, therefore, the sky in these regions possesses a richness which could never be guessed at by naked-eye study alone. Sweeping over the area roughly marked by the stars α, δ, γ and β is especially worth while. Towards Cygnus the richness of the starry background as revealed by binoculars is indescribably beautiful.

Mythology and History

The Harp or Lyre is, according to the old Greek myth, that wonderful instrument invented by Hermes and given by Apollo to Orpheus, who charmed all living creatures and even inanimate nature with his playing. After Orpheus' death his lyre was translated to the heavens to remind men for all times of Orpheus and his music.

Objects of Interest

α *Lyrae*. Vega. Mag. 0·2, bluish-white. The second brightest star in the northern skies. Distance about 26 L.Y.; luminosity, 50☉; diameter, 2·5☉. It is towards this part of the star sphere that the sun is moving with a velocity of 13 m.p.s.

β *Lyrae*. Eclipsing variable, the whole variation visible to the naked eye: mag. 3·4–4·1, secondary minimum at 3·8; period, 12d. 21h. 45m. Mass of A, 21☉; diameter, 36☉. Mass of B, 10☉; diameter, 28☉. Distance apart, less than one-third earth to sun. β has three companion stars, too faint to be detected with binoculars.

ϵ *Lyrae*. One of the most striking systems in the heavens, the celebrated 'double double'. Acute naked eyesight shows ϵ to be double—mags. 4, 5; separation, 3′ 28″—but binoculars are needed to show this clearly. A small telescope reveals that both ϵ^1 and ϵ^2 are themselves double, their respective separations being 2″·9 and 2″·3.

δ *Lyrae*. Close naked-eye double. Mags. 4, 5, orange and white; separation, 12′ 30″.

ζ *Lyrae*. Beautiful pair. Mag. 4 reddish, mag. 5·5 blue-green; separation, 44″.

Lyrids. April 20th–22nd. Meteors typically white and swift, and leave streaks; 8–15 per hour at maximum (April 21st). Odd Lyrids occur throughout May and June. Probably connected with Thatcher's comet (1861). Records of Lyrid showers go back to 687 B.C.

APPENDIX 3

SAGITTARIUS: The Archer

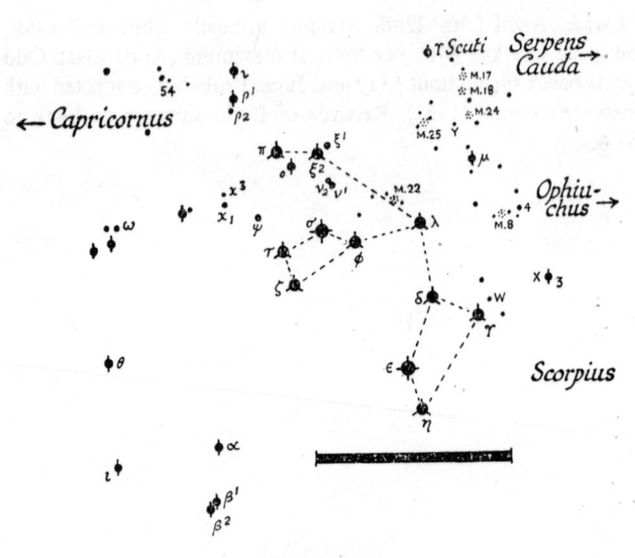

Figure 59

Position

A summer constellation, culminating at midnight early in July. It is a large constellation lying entirely south of the celestial equator; the Milky Way straggles across its western reaches. It is bounded on the north by Aquila, Scutum and Serpens Cauda, on the east by Capricornus, and on the west by Scorpius and Ophiuchus. The southernmost limits of the constellation are invisible in our latitudes.

Description

The Archer is a conspicuous constellation, and the ninth sign of the zodiac. The brighter stars fall into two quadrilaterals, with three 3rd mag. stars to the north of these. The Arabs called the eastern quadrilateral 'the Returning Ostriches' and the western 'the Going Ostriches'—going to and fro to drink from the Milky Way. There is fine sweeping in Sagittarius, especially over the dense star clouds that constitute the Milky Way in this region. It contains many fine clusters and galactic nebulae, some of which are mentioned below. Others which may be swept up are marked on the map.

Mythology and History

Old star atlases depict the Archer as standing with bow fully drawn, the arrow pointing at the heart of the Scorpion (*q.v.*). It was venerated in China at least 3,000 years ago, and was known to the ancient Babylonians and Persians. The Indians likened it to a horse 5,000 years ago. The classical tradition is that Chieron invented the constellation as a navigational guide for the Argonauts. Since Taurus sets as Sagittarius rises, the latter constellation was sometimes known as the Bull Killer.

Objects of Interest

54 Sagittarii. A faint double. Mags. 6, 7·5; separation, 45".

W Sagittarii. Cepheid; period, 7d. 14h. 10m. Mag. 4·8–5·8.

X (3) Sagittarii. Cepheid; period, 7d. 15m. Mag. 4·4–5.

Y Sagittarii. Cepheid. Period shorter, and mag. lower, than the two preceding variables: 5d. 18h. 30m.; 5·4–6·5.

M. 8, N.G.C. 6523. Fine nebulous cluster, visible to the naked eye. Appears as a loose cluster of rather faint stars, somewhat resembling a miniature Pleiades, the nebulosity not being visible in binoculars; even the individual stars may not be seen, the object then appearing as a structureless patch of misty light. Fine sweeping in this region; cluster prettily placed between two 5·5 mag. stars. Estimated distance, 1,600 L.Y.

M. 22, N.G.C. 6656. Fine globular cluster, just visible to the naked eye, midway between μ and σ. Unusually large and bright for this type of object, and only its low declination prevents it rivalling the Hercules cluster. The 50,000-odd stars are of mags. 10–15; resolution impossible with binoculars. Distance. about 10,000 L.Y. Discovered 1665.

M. 24, N.G.C. 6603. Magnificent star cloud, easily found about 3° north and slightly east of μ Sagittarii. To the naked eye, a protuberance from the Milky Way; with binoculars, an ill defined luminous patch. Fine sweeping all over this very rich region of the Galaxy.

AQUILA: THE EAGLE

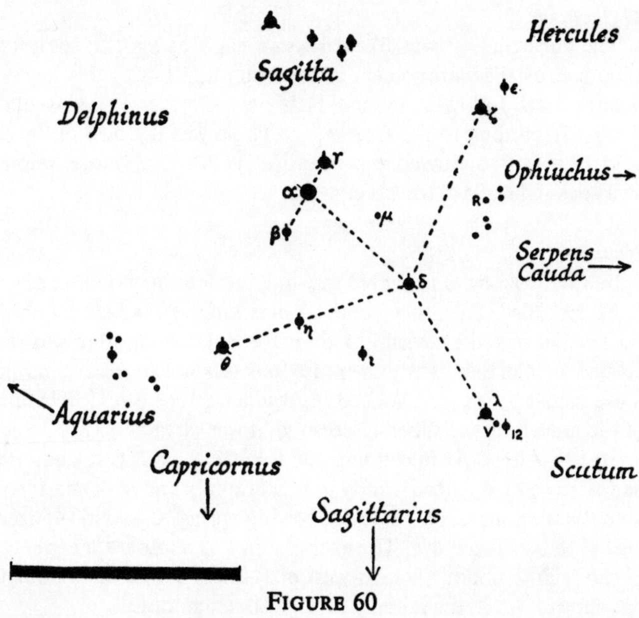

FIGURE 60

Position

A spring to autumn constellation, culminating at midnight in mid-July, which lies in and around the Milky Way south of Cygnus. α and its two attendant stars lie due south of β Cygni, the foot of the Cygnus Cross. The celestial equator runs through the constellation, Altair (α) lying some 8° north of it. To the south lie Capricornus and Sagittarius.

Description

Altair (mag. 1), flanked by the 3rd and 4th magnitude γ and β, is the chief naked-eye feature of the constellation, and renders its identification easy. To the south, and on the produced line through these three stars, lies θ, to the east of which is a curved line of faint stars, resembling a button-hook, which constitutes the obsolete constellation of Antinous.

Mythology

The constellation was likened to an eagle at least as early as 1200 B.C. by the astronomers of the Euphrates basin; this identification was followed by the Hebrews, Greeks, Romans and Arabs. According to the Greeks, the Eagle was the bird of Zeus, who bore off to heaven the beautiful youth Ganymede whom Zeus desired to have for his cup-bearer.

History

In A.D. 389 a nova appeared in Aquila which was bright enough to be recorded. Rivalling Venus at maximum, it had sunk below the level of naked-eye visibility three weeks later. Another was recorded by Chinese astronomers in 1609. Of novae discovered in more recent times, that noticed in Aquila on June 8th 1918 is one of the most famous. When discovered, it was already mag. 0·7 and a rival to Altair. At maximum, the following day, its magnitude had risen to –1·4; subsequently it faded rapidly and with considerable fluctuations until, by the following spring, it was no longer visible to the naked eye. The examination of photographic plates of the region (about 8° north-west of λ) taken before the outburst has shown that it was initially of the 11th magnitude.

Objects of Interest

α Aquilae. Altair, mag. 0·9. One of the sun's nearest neighbours, being only 16 L.Y. distant; its radial velocity is –20 m.p.s. Its size is estimated to be 1·5 ⊙, mass 5 ⊙ and luminosity 8 ⊙.

η Aquilae. Cepheid variable, totally visible to the naked eye. Mag. 3·7–4·5. Period, 7d. 4h. 14m., the rise to maximum taking 57 hours and the fall to minimum 115 hours. Distance, 650 L.Y. Radial velocity, –9 m.p.s.

R Aquilae. L.P.V., visible with binoculars near maximum. Mag. 5·8–11·8. Period, 310d.

V Aquilae. A variable worth looking up on account of its intense red colour. Mag. 6·5 at maximum, i.e. just on the threshold of naked-eye visibility.

APPENDIX 3

SAGITTA: The Arrow

Figure 61

Position

A summer constellation, culminating at midnight towards the end of June. Owing to the faintness of its stars it is useless to attempt to find Sagitta until the circumambient constellations—Aquila and Cygnus—are known. It lies in the Milky Way midway between the bright Aquila triplet (α, β and γ) and β Cygni, the foot of the Cygnus Cross.

Description

A small asterism of five faint stars arranged in the form of an arrow, only one of which is brighter than the 4th mag. Sweeping in this region reveals some fine star fields, notably in the region of η Sagittae and a few degrees south of γ.

Mythology and History

Despite its insignificance, Sagitta is one of the forty-eight constellations of the ancients. It was known as the Arrow to the Hebrews, Persians, Arabs and Romans. One legend states that it is the weapon with which Hercules slew the vulture which daily through the ages had torn and lacerated the vitals of Prometheus, chained to a rock face in the Caucasus Mountains. Another tradition says that it was the arrow with which Apollo killed Cyclops. Inevitably, too, it has been identified with Cupid's arrow.

Objects of Interest

ε *Sagittae.* Mags. 6, 7·8; separation, 92″. Within reach of an experienced eye and good binoculars.

α and β *Sagittae.* Both wide doubles.

APPENDIX 3

VULPECULA: The Fox

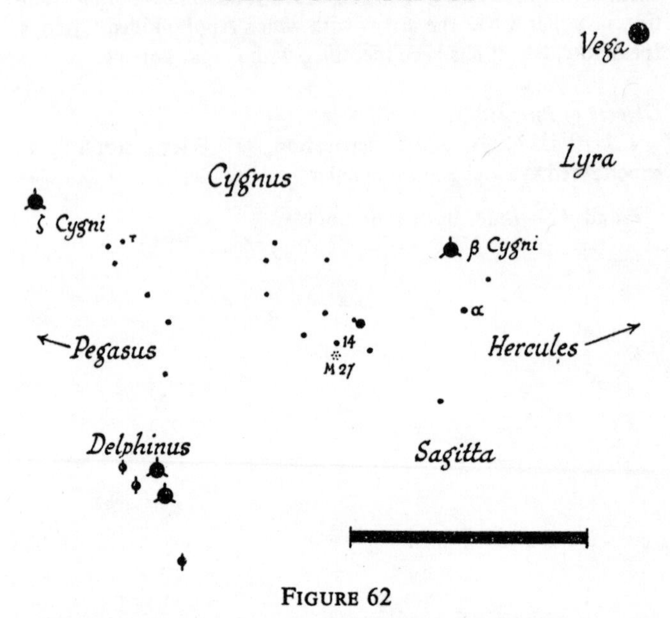

Figure 62

Position and Description

Vulpecula is a small and undistinguished constellation lying athwart the Milky Way due south of Cygnus. Hercules lies to its west, Lyra to its north-west, Pegasus to the east, and Delphinus and Sagitta to the south. It is thus a summer constellation, and culminates at midnight towards the end of July. Vulpecula is difficult to identify, possessing no stars brighter than magnitude 5. The greater part of the constellation lies within the triangle formed by the stars ζ Cygni, β Cygni and the asterism Delphinus.

Historical

Vulpecula was one of the new constellations created by Hevelius in the seventeenth century.

Objects of Interest

T Vulpeculae. Cepheid, period 4·436d. Mag. 5·2–6·4.

M. 27, N.G.C. 6853. The 'Dumb-bell Nebula', ½° S. of the mag. 5 star 14 Vulpeculae. It measures about 7′ by 9′. Appears as a small, featureless, nebulous spot in binoculars. Estimated distance, 550 L.Y.

APPENDIX 3

DELPHINUS AND EQUULEUS

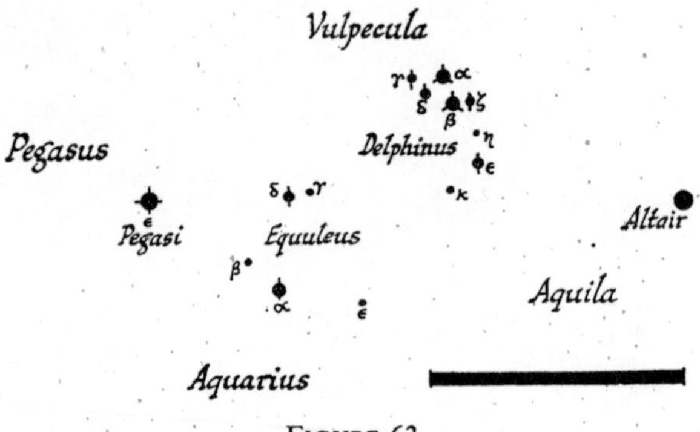

FIGURE 63

DELPHINUS: THE DOLPHIN

Position

A summer constellation culminating at midnight at the end of July. Since it is so small and comparatively inconspicuous it is inadvisable to look for it without first identifying Cygnus and Aquila. It lies 15° north-east of Altair (α Aquilae) and due south of Cygnus.

Description

Delphinus is a small, kite-shaped group of inconspicuous stars, two of the principal ones being mag. 3, and three being mag. 4. In size and general characteristics it resembles the nearby constellation Sagitta. α and β Delphini are known respectively as Sualocin and Rotanev, which spell backwards the names of the assistant of the eminent eighteenth to nineteenth-century astronomer Piazzi—Nicolaus Venator. This area of the sky, though uninteresting to the naked eye, repays sweeping over with binoculars.

Mythology and History

Both the Romans and Greeks called this constellation the Dolphin. Ovid called it Amphitrite, whom a dolphin persuaded to become Neptune's wife. To the Hindus it was known as the Porpoise. An alternative modern name for the asterism is Job's Coffin. Another old myth relates that it is the dolphin on whose back Arion, the Lesbian poet and musician, was carried to safety out of the hands of sailors who, envious of his skill, were planning to kill him. Neptune, in recognition of this service, translated the dolphin to the skies.

APPENDIX 3

EQUULEUS: THE LITTLE HORSE

One of the smallest of the constellations. It lies closely south-east of Delphinus, and contains nothing of interest for the observer with binoculars. Approximately midway between ϵ Pegasi and ϵ Delphini lie the two stars δ and γ Equulei (mags. 4·5, 5·0), some 7° to the south of which is the 4th mag. α. Mythologically it is thought to represent Cyllarus, the steed given to Pollux by Juno.

CYGNUS: THE SWAN

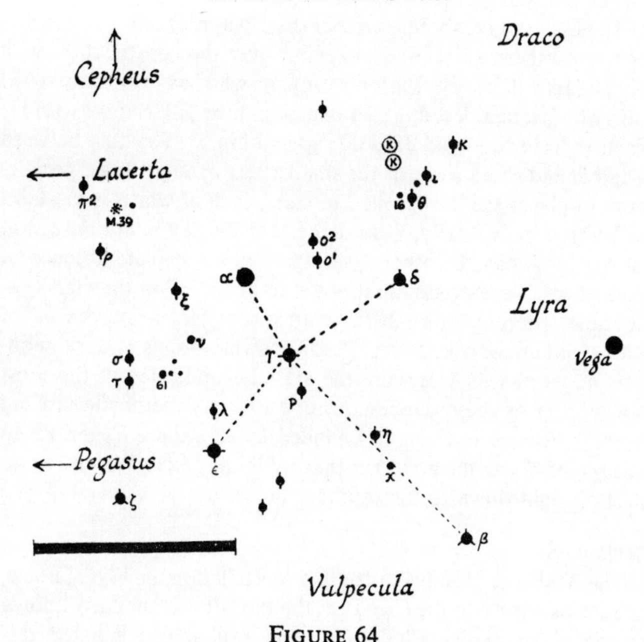

FIGURE 64

Position

A summer constellation, culminating at midnight at the end of July, when it is almost directly overhead. It is situated between Lyra (to the west) and the Great Square of Pegasus (to the east). During the autumn and early winter, though nearer the (western) horizon than during the summer, the Cross is upright in the sky instead of on its side, and is a fine spectacle.

Description

Its cruciform outline renders it easy to identify, although none of its stars is outstandingly brilliant. α is mag. 1·3, but of the four other stars marking the arms of the cross and their point of intersection, two are mag. 2 and the others mag. 3. β Cygni, mag. 3·2, marks the foot of the cross.

The Galaxy hereabouts is richer than in almost any other region of the northern skies, and sweeping over the constellation with binoculars will be a revelation to anyone who has previously relied solely on his unaided sight. In one area, near γ, Herschel with his great reflector counted 331,000 stars within 5°. The area between γ and β, and south towards the small asterism Sagitta, is especially rich. In places the background of stars, each of which is too faint to be seen individually, is so dense that the sky is not dark, but faintly luminous. In other areas there is a complete absence of stars. These were originally thought to be 'holes' in the Galaxy—i.e. lanes running through the star system which are devoid of stars—but are now known to be due to vast clouds of dark nebulous material which obscure the stars beyond. One of the most noteworthy of these dark nebulae, known as the northern Coal Sack, is situated in the area bounded by α, γ and ε Cygni. From this region starts the great rift that splits the Milky Way into two streams right down to the southern horizon.

Mythology

The Arab astronomers called this constellation the Flying Eagle, and it was known to the Greeks as the Bird. It was similarly known to the much earlier Chaldeans, who probably originated this identification. There are several classical myths referring to this celestial swan. One supposes that it is the swan whose form Jupiter assumed in order to seduce Leda, wife of the king of Sparta. Another identifies it as Orpheus, who on his death was changed into a swan and placed in the skies near to his beloved lyre.

Objects of Interest

α Cygni. One of the most intrinsically luminous stars. Distance about 1,600 L.Y., radial velocity –3 m.p.s.

β Cygni. One of the most beautiful doubles in our skies. Mags. 3·2 deep yellow, 5·5 green; separation, 35″. Good binoculars held steady will split it, but the strikingly contrasted colours will not be obvious without a telescope. Radial velocity, –15 m.p.s.

o^2 Cygni. Wide triple: mags. 4, 5·5, 7·5; separation, 358″, 107″.

61 Cygni. The first star, after the sun, whose distance was measured. This feat was accomplished by Bessel in 1838, by the trigonometrical parallax method; the distance is 10 L.Y. The velocity of its proper motion is abnormally high—about 50 m.p.s., corresponding to an angular displacement of nearly 9′ per century. Radial velocity, –40 m.p.s. It is a true binary, though not visible as such with binoculars, with a period of some 1,000 years. Though only of mag. 5·5 it can be easily picked up by binoculars with the help of the map, and is worth finding for its historical interest.

16 Cygni. Inconspicuous double: mags. 5·1, 5·3; separation, 39″. The components have the same proper motion.

R Cygni. Red variable; mag. 6·0–13·9; period, 426d. Easily found at maximum close to θ Cygni. Radial velocity –21 m.p.s.

χ Cygni. L.P.V. of the Mira (ο Ceti) type. Mag. 4·2–13·7, i.e. 9,500 times brighter at maximum than at minimum. Period, 410d. Visible to the naked eye when near maximum.

M. 39, N.G.C. 7092. Fine open cluster, situated midway between α Cygni and α Lacertae; just visible to the naked eye when its position is known beforehand. Resolution into starry points is accomplished with binoculars.

κ Cygnids. January 17th; slow-moving meteors, often trained. A second shower from a near-by radiant occurs August 14th–17th. Both weak.

APPENDIX 3

CAPRICORNUS: THE SEA GOAT

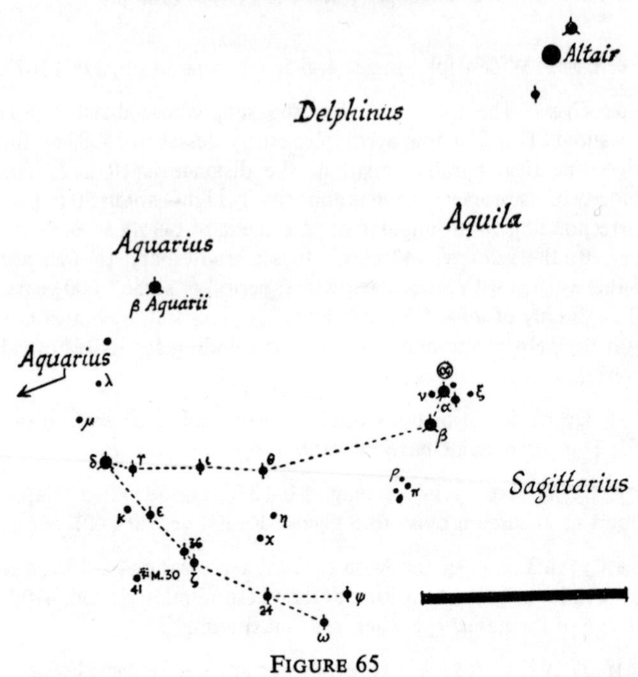

FIGURE 65

Position

Midnight culmination about August 8th. It lies entirely south of the equator, in a direct line with the three bright stars of Aquila, γ, α and β. It is some 20° due south of Delphinus and is bounded by the constellations Aquarius, Aquila, Sagittarius and the southern constellations Microscopium and Piscis Australis.

Description

An inconspicuous constellation to the casual glance, but easily enough found from the indications from Aquila and Delphinus. It has three 3rd mag. stars, the remainder being mag. 4 and lower. At each end of the constellation is a pair of relatively bright stars

—δ and γ to the east, and α and β to the west—which aid its identification. The tenth sign of the zodiac.

Mythology and History

It was named the Goat by the Babylonians and Chaldeans several millennia B.C., and in later times by the Greeks, Romans, Persians, Egyptians and Arabs. According to the Platonists it was the 'Gate of the Gods' through which the souls of men ascend to heaven. Old star atlases figure the constellation as a creature with the head of a goat and the tail of a fish.

Objects of Interest

α *Capricorni.* Naked-eye double, α^1 and α^2 (mags. 3, 4) being separated by 6′ 16″. Their similar yellow tint and almost equal brightness render them a striking pair in binoculars. α^1 has a companion star 45″ distant, and α^2 a mag. 11 comes which is itself a close double. α^1 is estimated to be 1,600 L.Y. distant; radial velocity, –16 m.p.s. α^2 is much nearer the sun—about 108 L.Y.

β *Capricorni.* Mags. 3·3 orange, 6·2 blue; separation, 205″. Easy with binoculars. The primary is itself a close binary. Distance about 465 L.Y.

M. 30, N.G.C. 7099. Globular cluster, about 3′ in diameter, near 41 Capricorni (mag. 5·5). Tiny nebulous spot in binoculars.

α *Capricornids.* July 23rd–31st; maximum, about July 29th. Connected with comet 1881 V. Weak shower of long, bright and slow moving meteors.

APPENDIX 3

AQUARIUS: THE WATER CARRIER

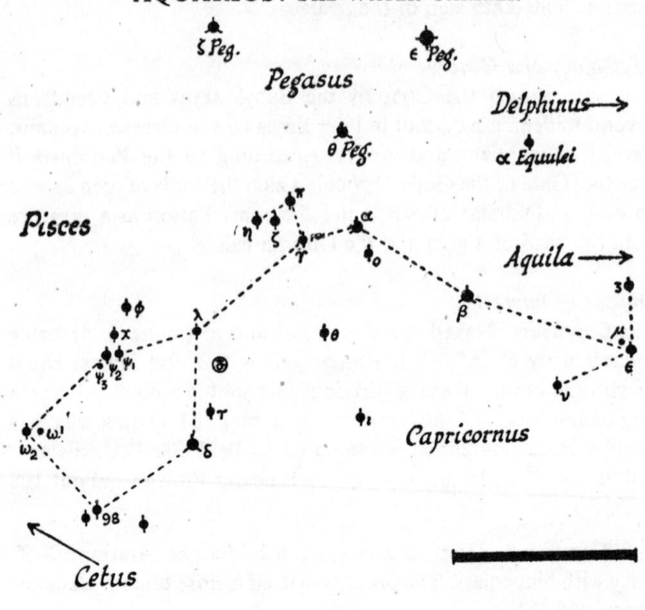

FIGURE 66

Position

A summer and autumn constellation, culminating at midnight around August 25th. It lies on the celestial equator (which passes between the stars π and ζ) south-west of the Great Square of Pegasus; that part of the constellation known as the 'Water Jar' is due south of ζ and θ Pegasi.

Description

The most noticeable feature of the outline of this constellation is the 'Water Jar', consisting of the stars π, η, ζ and γ. Westward from γ lie the stars α, β, υ, ϵ and 3 Aquarii, which somewhat resemble a distorted Plough. South and east from the 'Water Jar' the host of faint stars which the ancients used to call 'the Sea' stream down the sky towards Fomalhaut and the southern horizon. Aquarius, the eleventh sign of the zodiac, contains two fine planetary nebulae and many doubles, triples and groupings of faint stars, but these are for the most part inaccessible to binoculars. The area of 'the Sea' is worth sweeping over with glasses.

Mythology and History

From the earliest times this constellation has been associated with water, probably because the sun is to be found in it during the rainy month of February. The Babylonians saw it as a man pouring water out of a jar carried on his shoulder; the Arabs, a man carrying two water barrels; the Greeks called it Deucalion, or the water pourer; the Egyptians accounted for the annual flooding of the Nile by supposing that the water carrier dipped his gigantic ewer into the springs of the river to replenish it once a year.

Turning from mythology to history, it was within 1° of ι Aquarii that Galle at Berlin discovered the planet Neptune on 23rd September 1846, close to the position predicted independently by Adams and Leverrier.

Objects of Interest

γ *Aquarii.* Mag. 4 star of a distinct green tint. Fine sweeping southwards to the horizon.

λ *Aquarii.* Unusually red star.

τ *Aquarii.* A fine double in binoculars. White and reddish.

δ *Aquarids.* July 20th–August 4th, maximum about July 28th. Meteors typically long and slow moving. Rather far south for effective observation in the British Isles.

APPENDIX 3

PISCIS AUSTRALIS: THE SOUTHERN FISH

Position

As its name implies, this constellation is situated far south in our skies, α Piscis Australis being exactly 30° south of the celestial equator. The constellation lies to the south of Aquarius and Capricornus, and Fomalhaut (being the only bright star in this part of the sky) is easily identified. It culminates at midnight during August.

Description

Fomalhaut (see map of Aquarius, p. 240) is really the only object in this constellation which concerns observers in the British Isles, the remainder of Piscis being below the horizon.

Mythology and History

Piscis Australis is one of the original forty-eight constellations of the ancients, and 5,000 years ago α was one of the four 'Royal Stars' of the Persians. Fomalhaut is an Arab word meaning 'Mouth of the Fish'.

Objects of Interest

α Piscis Australis. Fomalhaut. One of the few bright stars of a markedly red tint. Mag. 1·3; distance, 23 L.Y.; radial velocity, +4 m.p.s.; proper motion, 1° per 6,000 years. It has two telescopic companions at distances of 30″ and 72″.

PEGASUS: The Winged Horse

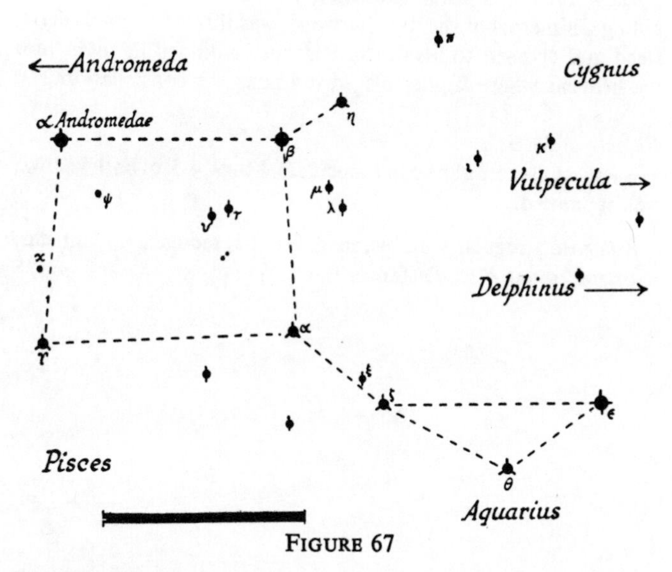

FIGURE 67

Position

An autumn constellation, culminating at midnight early in September. Pegasus covers a large area, lying roughly between Delphinus and Cygnus to the west, Aquarius to the south, Pisces and Andromeda to the east, and Lacerta to the north. Easily found from its proximity with Andromeda (q.v.). The Great Square of Pegasus is one of the 'landmarks' of the night sky.

Description

Delineations of the Winged Horse have been found on very early ceramics and tablets from the Euphrates valley. At a later date (*c.* 500–100 B.C.) it appears on coins minted at Corinth. The Pegasus myth runs very briefly as follows: Pegasus sprang from the spilt blood of the decapitated Medusa, slain by Perseus, and immediately flew off into the sky, finally returning to earth on the summit of Mt. Helicon. It was eventually tamed by Minerva or Neptune, and presented to Bellerophon in order to carry him to Lycia where the monster Chimera lived. Bellerophon succeeded in killing Chimera, but shortly afterwards was thrown from his aerial steed and crashed to his death. Pegasus continued its flight into the heavens where Jupiter placed it among the constellations.

Objects of Interest

π Pegasi. Beautiful pair in binoculars. Mags. 4·5, 6, both yellow; well separated.

β Pegasi. Irregular variable, mag. 2·2–2·8, reddish. A giant star, diameter about 87⊙. Distance, 196 L.Y.

LACERTA: THE LIZARD

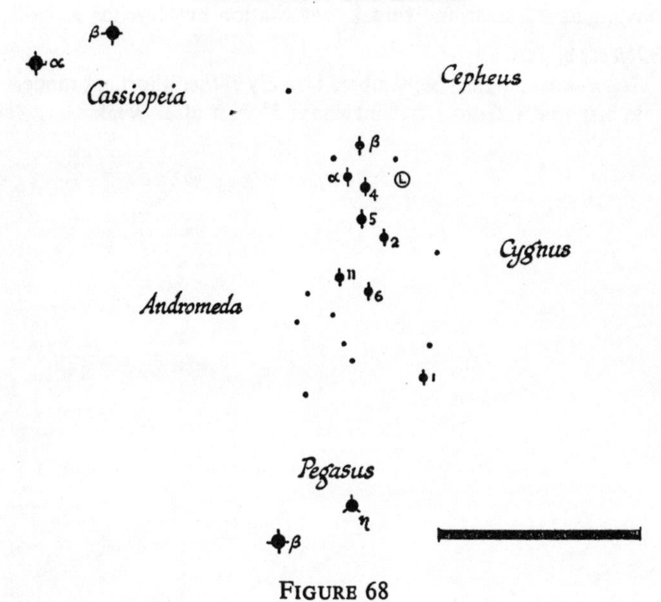

FIGURE 68

Position

Lacerta is partly circumpolar, culminating at midnight during September. It is sandwiched between Andromeda on the east and Cygnus on the west; to the south lies Pegasus, to the north Cepheus. The Great Square lies closely south-east, the W of Cassiopeia being about the same distance north-east.

Description

This is a small constellation of faint stars, the eight brightest all being of the 4th mag. With the help of the accompanying map it should be easily enough identified.

APPENDIX 3

History

The Lizard is one of the more recent constellations, having been distinguished as an individual constellation by Hevelius *c.* 1690.

Objects of Interest

Lacertids. August–September. Usually rather short, of moderate brightness, from a radiant about 3° west of α. Weak.

CEPHEUS

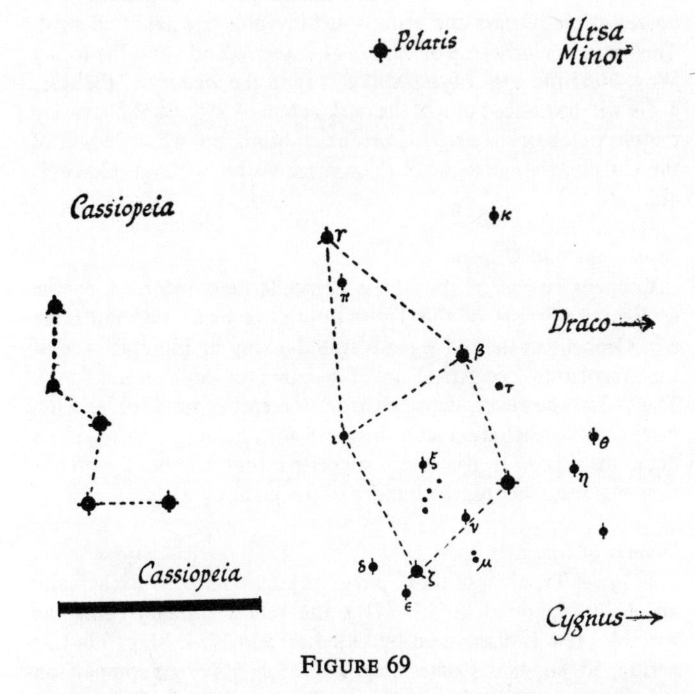

FIGURE 69

Position

A circumpolar constellation lying for the most part between the Milky Way and Polaris in the triangular area between the three constellations Cygnus, Cassiopeia and Ursa Minor. The two stars α–β Cassiopeiae point direct at α Cephei, some 20° from β Cassiopeiae. Cepheus culminates at midnight early in September.

Description

An inconspicuous constellation, the brightest star (α) being mag. 2, and the three next brightest mag. 3. The outline marked by γ, ι, β, ζ and α resembles that of a church spire, the tip of which points roughly in the direction of Polaris. Though uninteresting to the naked eye, Cepheus repays examination with glasses for it possesses many stars and groups just invisible to unassisted sight. This is particularly true of the area between α and ι and the Milky Way. Near the edge of the Milky Way in the vicinity of the stars δ–ζ–ϵ will be noticed one of the dark nebulae—clouds of obscuring matter which occur in great profusion along the whole length of the Galaxy, though most of them are invisible without telescopic aid.

Mythology and History

Cepheus is one of the oldest constellations, recorded by the astronomer-priests of the Tigris-Euphrates basin two millennia B.C. Cepheus in the old legends was the king of Ethiopia, one of the Argonauts (see Aries) and the father of Andromeda (q.v.). The Arabs, however, departed from the earlier tradition and depicted the constellation as a shepherd with his dog overlooking a flock of sheep. It has been suggested that to the Egyptians, Cepheus was Cheops, the builder of the great pyramid.

Objects of Interest

δ Cephei. Type star of the Cepheid variables (see p. 88) δ is actually an easily separated double (41″), the yellow primary being the variable; it was discovered by Goodricke in 1784. Mag. 3·6–4·3; period, 5d. 8h. 48m. Comes mag. 7·5 bluish. Having a comparison star so near makes the fluctuating brightness of δ all the more easily observed. Distance about 295 L.Y.

β Cephei. A double whose components are unfortunately too close (14″) for separation with binoculars. It is interesting, nevertheless, for the fact that the primary is itself a sp. bin. whose period of 5h. is one of the shortest known. Radial velocity, –9 m.p.s.

μ Cephei. A variable of unusually deep red colour, more noticeable with binoculars or a small telescope than with the naked eye; comparison with the white α Cephei emphasizes its red tint. μ

varies irregularly between mags. 3·7 and 4·7 in a period of 5–6 years.

T Cephei. L.P.V., just visible to the naked eye when near maximum. Mag. 5·2–9·5, period, 390d.

RZ Cephei. A faint variable worth mentioning since its radial velocity is one of the highest known—nearly 700 m.p.s.

GLOSSARY OF TERMS AND ABBREVIATIONS

Achromatism. Freedom from chromatic aberration (q.v.); the focusing of light of different colours (wavelengths) to the same point.

Ångstrom Unit (*Å*). The unit of wavelength employed with very short-wave radiation; its value is 10^{-8} cms (one hundred-millionth of a centimetre).

Annular. Ring-like; solar eclipses in which a ring of the sun's photosphere is visible round the limb of the moon are so called.

Aphelion. That point on the orbit of a planet or comet at which it is farthest from the sun.

Astronomical Unit (*A.U.*). The earth's mean solar distance: 93 million miles.

Chromatic aberration. An optical fault in simple lenses whereby they focus light of different colours (wavelengths) at different distances from the lens.

Circumpolar stars. Those which can never set, since their angular distance from the celestial pole is less than that of the pole from the horizon at the place of observation.

Collimator. The optical part which makes the light rays parallel before they strike the prism or diffraction grating of a spectroscope.

Comes. The fainter member of a double star system.

Conjunction. One or more planets and the sun or moon are said to be in conjunction with one another when they lie on or close to the same line of sight from the earth.

Constellation. A grouping of stars, usually more or less conspicuous, together with the area of the star sphere covered by it.

d.h.m. Days, hours, minutes.

GLOSSARY OF TERMS AND ABBREVIATIONS

Declination. The system of co-ordinates on the star sphere corresponding to latitude on the earth. The Dec. of the celestial equator is 0°, and that of the celestial poles ±90°.

Dispersion. The angular separation of different wavelengths by a spectroscope.

Diurnal. 24-hourly.

Ecliptic. The path traced out on the star sphere by the centre of the sun's disc during its annual circuit of the heavens. Alternatively, the line of intersection of the produced plane of the earth's orbit with the star sphere.

Ellipse. A closed oval figure, exemplified by the planetary orbits; one of the conic sections, whose eccentricity is less than unity.

Elongation, maximum. The greatest angular distance to which an inner planet recedes from the sun, or a satellite from its planet.

Equator, celestial. The great circle on the star sphere which is the projection of the earth's equatorial plane; it cuts the horizon at the east and west points.

Equinox. The two intersecting points of the ecliptic and the celestial equator. The sun passes these points on about March 21st (Spring Equinox) and September 21st (Autumn Equinox), when the day and night are of equal length.

Focal length. The distance from the centre of a lens or mirror to the point at which it focuses parallel incident light.

Focal ratio. The ratio between the diameter and focal length of a lens or mirror; e.g. f/6 means that the focal length is 6 times the aperture.

Galaxy. The star system of which the sun is a member; or, the Milky Way; or, any other star system or extragalactic nebula.

Greek alphabet. See. p. 132.

Ionization. The removal of one or more of the electrons circling round an atomic nucleus.

L.P.V. Long period variable.

Luminosity. The intrinsic brightness of a star.

L.Y. Light year: the distance travelled by a ray of light in one year (nearly six billion miles).

Mag. Stellar magnitude; the unit of apparent brightness of stars and other celestial objects.

Meridian. The great circle which passes through the zenith, the celestial poles, and the north and south points of the horizon; or, any great circle on the surface of the earth, sun, or other body, which passes through its poles—e.g. the Greenwich meridian, the sun's central meridian, etc.

Monochromatic. Of a single wavelength.

Objective. The primary focusing unit in a telescope or camera, whether lens (object glass) or mirror.

Obliquity of the ecliptic. The angle between the earth's equatorial plane and the plane of its orbit = the angle on the star sphere between the celestial equator and the ecliptic = $23\frac{1}{2}°$.

Occultation. The passage of one celestial body behind a nearer one, as seen by a terrestrial observer.

Off-axis aberrations. Defects in optical systems which result in the distortion of images in proportion to their distance from the centre of the field, i.e. they progressively affect an image-forming ray as its inclination to the axis of the instrument is increased.

Opposition. Any two celestial bodies are said to be in opposition when they are diametrically opposite one another on the star sphere.

Parabola. A conic section whose eccentricity is equal to unity. A surface, all of whose median sections are parabolas, is called a paraboloid; this is the figure of telescopic mirrors.

Parallax. The apparent displacement of a distant object resulting from a change in the position of the observer.

Perihelion. That point on the orbit of a planet or comet at which it is nearest to the sun.

Polychromatic. Radiation consisting of many wavelengths, e.g. white light.

Primary. The brighter member of a double star system.

Proper motion. The component of a star's motion which lies in the plane at right angles to the observer's line of sight.

Radial velocity. The velocity of a star or other body in the observer's line of sight.

Radiant. The point on the star sphere from which the individual members of a meteor shower appear to radiate.

Refraction. The deflection of a light ray from its original course on passing from one optical medium to another, e.g. through a lens or prism (air—glass—air).

Resolution, resolving power. The capacity of the eye or telescope to separate the images of objects which are angularly close to one another.

Right Ascension (*R.A.*). The system of co-ordinates on the star sphere corresponding to longitude on the earth. Measured in hours from 0 to 24, starting from the 'First Point of Aries'—the Spring Equinox.

Solstice. The sun is at the Summer Solstice when farthest north of the celestial equator (about June 21st), and at the Winter Solstice when farthest south of it (about December 21st).

Sp. bin. Spectroscopic binary.

'*Speed*' (*photographic*). A measure of the length of exposure required by a camera lens to form an image of a given density under standard conditions. For extended images it is a function of the focal ratio (q.v.); for point images such as those of stars, of the aperture of the lens only.

Spherical aberration. A defect of spherical mirrors whereby rays reflected from the mirror's edge are brought to a shorter focus than rays reflected from the centre.

Star sphere. The illusory sphere, popularly called the dome of heaven, to which the celestial bodies appear to be attached.

Terminator. The demarcation line on the lunar surface which separates the sunlit and the dark hemispheres.

Transit. The passage, as seen from the earth, of one celestial body in front of a more distant one.

Wavelength. The distance separating adjacent crests or adjacent troughs in a system of waves.

Zenith. The point on the star sphere which is vertically above the observer.

Zodiac. An 18°-wide zone on the star sphere within which the ecliptic is centrally placed.

⊙ The sun. Hence 'Mass 175⊙' means 'mass is 175 times that of the sun'.

° ′ ″. Degrees, minutes, seconds.

INDEX

INDEX

INDEX